587200

ONE WEEK LOAN

009404024

NAVIGATIONAL SYSTEMS AND SIMULATORS

Navigational Systems and Simulators

Marine Navigation and Safety of Sea Transportation

Editor

Adam Weintrit
Gdynia Maritime University, Gdynia, Poland

CRC Press
Taylor & Francis Group
Boca Raton   London   New York   Leiden

CRC Press is an imprint of the
Taylor & Francis Group, an **informa** business

A BALKEMA BOOK

CRC Press/Balkema is an imprint of the Taylor & Francis Group, an informa business

© 2011 Taylor & Francis Group, London, UK

Printed and bound in Great Britain by Antony Rowe Ltd (A CPI-group Company), Chippenham, Wiltshire

All rights reserved. No part of this publication or the information contained herein may be reproduced, stored in a retrieval system, or transmitted in any form or by any means, electronic, mechanical, by photocopying, recording or otherwise, without written prior permission from the publisher.

Although all care is taken to ensure integrity and the quality of this publication and the information herein, no responsibility is assumed by the publishers nor the author for any damage to the property or persons as a result of operation or use of this publication and/or the information contained herein.

Published by: CRC Press/Balkema
P.O. Box 447, 2300 AK Leiden, The Netherlands
e-mail: Pub.NL@taylorandfrancis.com
www.crcpress.com – www.taylorandfrancis.co.uk – www.balkema.nl

ISBN: 978-0-415-69113-0 (Pbk)
ISBN: 978-0-203-15734-3 (eBook)

List of reviewers

Prof. Michael **Barnett**, Southampton Solent University, United Kingdom,

Prof. Eugen **Barsan**, Master Mariner, Constanta Maritime University, Romania,

Prof. Knud **Benedict**, University of Wismar, University of Technology, Business and Design, Germany,

Prof. Tor Einar **Berg**, Norwegian Marine Technology Research Institute, Trondheim, Norway,

Prof. Carmine Giuseppe **Biancardi**, The University of Naples „Parthenope", Naples, Italy,

Prof. Jarosław **Bosy**, Wroclaw University of Environmental and Life Sciences, Wroclaw, Poland,

Sr. Jesus **Carbajosa Menendez**, President of Spanish Institute of Navigation, Spain,

Prof. Shyy Woei **Chang**, National Kaohsiung Marine University, Taiwan,

Prof. Jerzy **Czajkowski**, Gdynia Maritime University, Poland,

Prof. Krzysztof **Czaplewski**, Polish Naval Academy, Gdynia, Poland,

Prof. Eamonn **Doyle**, National Maritime College of Ireland, Cork Institute of Technology, Cork, Ireland,

Prof. Andrzej **Fellner**, Silesian University of Technology, Katowice, Poland,

Prof. Andrzej **Felski**, President of Polish Navigation Forum, Polish Naval Academy, Gdynia, Poland,

Prof. Wlodzimierz **Filipowicz**, Master Mariner, Gdynia Maritime University, Poland,

Prof. Börje **Forssell**, Norwegian University of Science and Technology, Trondheim, Norway,

Prof. Wieslaw **Galor**, Maritime University of Szczecin, Poland,

Prof. Lucjan **Gucma**, Maritime University of Szczecin, Poland,

Prof. Stanisław **Gucma**, Master Mariner, President of Maritime University of Szczecin, Poland,

Prof. Jerzy **Hajduk**, Master Mariner, Maritime University of Szczecin, Poland,

Prof. Michal **Holec**, Gdynia Maritime University, Poland,

Prof. Jacek **Januszewski**, Gdynia Maritime University, Poland,

Prof. Tae-Gweon **Jeong**, Master Mariner, Secretary General, Korean Institute of Navigation and Port Research,

Prof. Mirosław **Jurdzinski**, Master Mariner, FNI, Gdynia Maritime University, Poland,

Prof. John **Kemp**, Royal Institute of Navigation, London, UK,

Prof. Serdjo **Kos**, FRIN, University of Rijeka, Croatia,

Prof. Pentti **Kujala**, Helsinki University of Technology, Helsinki, Finland,

Dr. Dariusz **Lapucha**, Fugro Fugro Chance Inc., Lafayette, Louisiana, United States,

Prof. David **Last**, FIET, FRIN, Royal Institute of Navigation, United Kingdom,

Prof. Joong Woo **Lee**, Korean Institute of Navigation and Port Research, Pusan, Korea,

Prof. Andrzej S. **Lenart**, Gdynia Maritime University, Poland,

Prof. Józef **Lisowski**, Gdynia Maritime University, Poland,

Prof. Vladimir **Loginovsky**, Admiral Makarov State Maritime Academy, St. Petersburg, Russia,

Prof. Evgeniy **Lushnikov**, Maritime University of Szczecin, Poland,

Prof. Aleksey **Marchenko**, University Centre in Svalbard, Norway,

Prof. Francesc Xavier **Martinez de Oses**, Polytechnical University of Catalonia, Barcelona, Spain,

Prof. Torgeir **Moan**, Norwegian University of Science and Technology, Trondheim, Norway,

Prof. Wacław **Morgas**, Polish Naval Academy, Gdynia, Poland,

Prof. Reinhard **Mueller**, Master Mariner, Chairman of the DGON Maritime Commission, Germany,

Prof. Janusz **Narkiewicz**, Warsaw University of Technology, Poland,

Prof. Andy **Norris**, FRIN, CNI, The Royal Institute of Navigation, University of Nottingham, UK,

Prof. Zbigniew **Pietrzykowski**, Maritime University of Szczecin, Poland,

Prof. Francisco **Piniella**, University of Cadiz, Spain,

Prof. Aydin **Salci**, Istanbul Technical University, Maritime Faculty, ITUMF, Istanbul, Turkey,

Prof. Jens-Uwe **Schroeder**, Master Mariner, World Maritime University, Malmoe, Sweden,

Prof. Roman **Smierzchalski**, Gdańsk University of Technology, Poland,

Prof. Henryk **Sniegocki**, Master Mariner, MNI, Gdynia Maritime University, Poland,

Prof. Jac **Spaans**, Netherlands Institute of Navigation, The Netherlands,

Prof. Cezary **Specht**, Polish Naval Academy, Gdynia, Poland,

Cmdr. Bengt **Stahl**, Nordic Institute of Navigation, Sweden,

Prof. Anna **Styszynska**, Gdynia Maritime University, Poland,

Prof. Marek **Szymoński**, Master Mariner, Polish Naval Academy, Gdynia, Poland,

Capt. Rein **van Gooswilligen**, Master Mariner, Chairman of EUGIN,

Prof. František **Vejražka**, FRIN, Czech Institute of Navigation, Czech Technical University in Prague, Czech,

Prof. Peter **Voersmann**, President of German Institute of Navigation DGON, Deutsche Gesellschaft für Ortung und Navigation, Germany,

Prof. Vladimir **Volkogon**, Rector of Baltic Fishing Fleet State Academy, Kaliningrad, Russia,

Prof. Jin **Wang**, Liverpool John Moores University, UK,

Prof. Ryszard **Wawruch**, Master Mariner, Gdynia Maritime University, Poland,
Prof. Bernard **Wisniewski**, Maritime University of Szczecin, Poland,
Prof. Jia-Jang **Wu**, National Kaohsiung Marine University, Kaohsiung, Taiwan (ROC),
Prof. Homayoun **Yousefi**, MNI, Chabahar Maritime University, Iran,
Prof. Wu **Zhaolin**, Dalian Maritime University, China

Contents

Navigational Systems and Simulators. Introduction

A. Weintrit

Gdynia Maritime University, Gdynia, Poland

PREFACE

The contents of the book are partitioned into six parts: global navigation satellite systems (covering the chapters 1 through 5), positioning systems (covering the chapters 6 through 11), navigational simulators (covering the chapters 12 through 20), radar and navigational equipments (covering the chapters 21 through 24), ship handling and ship manoeuvring (covering the chapters 25 through 26), search and rescue operations (covering the chapters 27 through 28).

The first part deals with global navigation satellite systems (GNSS). Certainly, this subject may be seen from different perspectives. The contents of the first part are partitioned into five chapters: A look at the development of GNSS capabilities over the next 10 years, GNSS meteorology, Onboard wave sensing with velocity information GPS, EGNOS performance improvement in Southern latitudes, and An integrated vessel tracking system by using AIS, Inmarsat and China Beidou Navigation Satellite System.

The second part deals with positioning systems. The principles of position, course and velocity determination are treated; radionavigation and terrestrial methods are presented. The contents of the second part are partitioned into six chapters: Recent advances in wide area real-time precise positioning, Assessing the limits of e-Loran positioning accuracy, Fuzzy evidence in terrestrial navigation, Ground-based, hyperbolic radiolocation system with spread spectrum signal – AEGIR, An algorithmic study on positioning and directional system by free gyros, and Compensation of magnetic compass deviation at one any course.

The third part deals with navigational simulators. Different kinds of navigational and manoeuvring simulators and simulating methods are presented. The contents of the third part are partitioned into nine chapters: New level of integrated simulation interfacing ship handling simulator with safety and security trainer (SST), Path following problem for a DP ship simulation model, Simulating method of ship's turning-basins designing, Capabilities of ship handling simulators to simulate shallow water, bank and canal effects, Development of a costs simulator to assess new maritime trade routes, Analogical manoeuvring simulator with remote pilot control for port design and operation improvement, Simulation model for detecting vessel conflicts within a seaport, Research on ship navigation in numerical simulation of weather and ocean in a bay, and A methodological framework for evaluating maritime simulation.

The fourth part deals with radar and navigational equipments. The contents of the fourth part are partitioned into four chapters: Impact of internal and external interferences on the performance of a FMCW radar, Fusion of data received from AIS and FMCW and pulse radar - results of performance tests conducted using hydrographical vessels "Tukana" and "Zodiak", Statistical analysis of simulated radar target's movement for the needs of multiple model tracking filter, and The modes of radar presentation of situation in inland navigation.

The fifth part deals with ship handling and ship manoeuvring. The contents of the fifth part are partitioned into two chapters: Multirole population of automated helmsmen in neuroevolutionary ship handling, and Ship's turning in the navigational practice.

The sixth part deals with search and rescue operations. The contents of the sixth part are partitioned into two chapters: Iridium a more effective proposal for the localization, search and rescue in the sea, and Research on the risk assessment of man overboard in the performance of Flag Vessel Fleet (FVF).

This book completes the body of the books series with outlook into the future of navigation and is mainly concerned with the scheduled provision of radio navigation systems, including GNSS in the near and medium-term future.

Global Navigation Satellite System

1. A Look at the Development of GNSS Capabilities Over the Next 10 Years

J. Januszewski

Gdynia Maritime University, Gdynia, Poland

ABSTRACT: This paper considers what the SNS (Satellite Navigation Systems) as GPS, GLONASS, Galileo and Compass, and SBAS (Satellite Based Navigation Systems) as EGNOS, WAAS, MSAS and GAGAN services might look like 10 years from now. All these systems, called GNSS (Global Satellite Navigation System), are undergoing construction or modernization (new satellites, new frequencies, new signals, new monitoring stations, etc.) and continuous improvement to increase its accuracy, availability, integrity, and resistance to interference. The most significant events in SNS and SBAS in the nearest 10 years are presented also. Additionally three possible scenarios considering these systems (in 2016 and 2021 years), concerning the number of satellites in particular, optimistic, pessimistic and the most probable were taken into account.

1 INTRODUCTION

Nowadays (January 2011) the American GPS Satellite Navigation System (SNS) is fully operational with 31 satellites. Since few years the Russian GLONASS system was being revamped and undergoing an extensive modernization effort, therefore today this system with 21 satellites can be used for fix position also. Galileo system (Europa) and Compass system (China) are under construction, must likely these systems will be operating at the earliest in 2016 and 2021 adequately.

The Satellite Based Augmentation Systems (SBAS) that enhance the integrity, accuracy, and operation of two SNS – GPS and GLONASS. Today the SBAS as Wide Area Augmentation System (WAAS), Multi-functional Transport Satellite Based Augmentation System (MSAS) and European Geostationary Navigation Overlay System (EGNOS) are accessible in USA and Canada, Japan and Europe and North Africa adequately. While WAAS and MSAS are fully operational since few years, EGNOS officially entered into operational phase with the provision of the Open Service as of only October 1, 2009. Additionally the Department of Defense of the United States is cooperating with India to develop new system over Indian space. This is the GAGAN (GPS and Geo Augmented Navigation), new SBAS, actually under construction. Other SBAS will enhance GLONASS and GPS systems, called SDCM (System for Differential Correction and Monitoring), is under construction in Russia. All these SNS and SBAS create Global Navigation Satellite System (GNSS). System Compass was nottaken into account in this paper, because about this system little information is available still.

2 SATELLITE NAVIGATION SYSTEMS CONSTELLATION

Actually (January 2011) GPS spatial segment consists of 32 satellites, 11 the oldest block IIa, 12 block IIR, 8 block IIR–M and 1 block IIF (Table 1) [www.navcen.uscg.gov]. Additionally in this table we can find the information about active life of each satellite in years and months. The value of this life depends on the system and satellite block. The mean values of these satellites life of block IIA, IIR and IIR–M are equal 16.3, 8.9 and 3.2 years adequately. It means that the active life of all satellites of block IIA is greater than nominal value 10 years, considerably. The satellites IIA and IIR transmit one frequency (L1) for civil users only, IIR–M two frequencies (L1, L2), IIF and the future III three (L1, L2, L5). Information about integrity will provide the satellites block III only [Gleason S., Gebre-Egziabher D. 2009] and [Hofmann-Wellenhof B. et all. 2008].

The GLONASS spatial segment consists of 21 satellites, all block M (Table 2), which transmit two frequencies for civil users (L1, L2), but without information about integrity [www.glonass-ianc.rsa.ru]. This information and the third frequency will be provided by the satellites next generation K. The Galileo spatial segment consists of 2 satellites only,

27 operational and 3 active spheres in the future. The satellites will transmit four frequencies.

The accuracy of the user's position obtained from the SNS depends on a number of satellites (ls) visible above masking angle. That's why the total number of satellites, fully operational especially, is very interesting for the users. There is no direct relation between the number ls and the position error M, but for all SNS in the case of position fix in restricted area we can say the following "when ls greater, M is less" and inversely "when ls is less, M is greater" [Januszewski J. 2008].

3 THE MOST SIGNIFICANT EVENTS IN THE GNSS IN THE NEAREST 10 YEARS

The most significant events in the SNS and SBAS waited in optimistic scenario into 10 nearest years (2012–2021) with the consequences for the civil users are presented in Table 3. One of the parameters mentioned in this table is the number of frequencies transmitted by the satellites of each SNS. Because of two or three frequencies make possible the calculation of ionosphere correction, the user's position accuracy increases. Unlike actual generation of GPS and GLONASS systems next generation of these systems, GPS III and GLONASS K, and new system Galileo will provide integrity information. Integrity can be defined as a reliability indicator of the quality of positioning, user's position obtained from SNS also [Januszewski J. 2009].

EGNOS has claimed that they will eventually transmit integrity information for users of GPS and GLONASS systems as well as for Galileo system.

Between 2008 and 2013, the FAA (Federal Aviation Administration) will make the necessary changes in the ground equipment of WAAS to handle the L5 signal from GPS. Having two frequencies for ionospheric corrections will eliminate loss of vertical guidance caused by ionospheric storms.

Japan has had a plan to display a new regional system called the Quasi–Zenith Satellite System (QZSS), which services include enhanced accuracy GPS signals, communications and broadcasting.

The GPS and GLONASS systems are undergoing uninterrupted modernization (new satellites, new frequencies, new signals, new codes, new monitoring stations, etc.) and continuous improvement to increase its accuracy (position in particular), availability, integrity, and resistance to interference, while at same time maintaining at least the performance it enjoys today with existing already user's receivers [Januszewski J. 2010] and [Springer T., Dach R. 2010]. In the case of the GPS system the plans of the control segment modernization are well known. The next Generation GPS Control Segment (OCX) will provide significant benefits to all users around the world, as well as to GPS operators, mainters, and analysts. Two major upgrades are in development; the Legacy Accuracy Improvement Initiative (L-AII) and the Architecture Evolution Plan (AEP). The L-AII upgrade adds up to 14, actually 11 only, National Geospatial Intelligence Agency (NGA) monitor stations [Kaplan E.D., Hegarty C.J. 2006], [Gower A. 2008].

United States Air Force officials are moving to reconfigure the GPS constellation to create as soon as possible a 27 satellites geometry that will improve the availability and accuracy of positioning, navigation, and timing capabilities, in particular for U.S. military forces [Roper E. 2010].

A third civil signal at the GLONASS L3 frequency will be on newer GLONASS K satellites, probably starting in 2011 (Table 3).

The first two in–orbit validation (IOV) Galileo satellites are scheduled for launch 2011, followed by two more in next year.

4 THE POSSIBLE SCENARIOS AFFECTING THE DEVELOPMENT OF GNSS

Three possible scenarios considering three SNS, the GPS, GLONASS and Galileo, and SBAS in 2016 and 2021 years, optimistic, pessimistic and the most probable were taken into account [Lavrakas J.W. 2007]. The projected total number of satellites, number of satellites transmitting signals for civil users on two and three frequencies and information about integrity for GPS, GLONASS and Galileo for each mentioned above scenario are presented in the author's Table 4.

4.1 *Optimistic scenario*

In this scenario every project meets its projected dates. In the case of GPS system the following assumptions are made for 2016 year:
- all 12 Block IIF and 4 Block III satellites were launched,
- as in 2011 the satellites IIA launched in 1992 or earlier are fully operational still, we can expect that in 2016 years the vitality of all satellites on orbit will be also 20 years.

In this situation we have in GPS satellites:
- 12 Block IIFs ranging from 0 to 6 years old,
- 8 Block IIR–Ms ranging from 7 to 11 years old,
- 12 Block IIRs ranging from 12 to 18 years old,
- 4 Block IIAs ranging from 19 to 20 years old,
- 4 Block IIIs ranging from 0 to 2 years old.

It means that the GPS spatial segment will consist of 40 satellites. As this number is greater than 32 (nominal value), 8 oldest satellites will be able to be not used. In 2016 year two other SNS the GLONASS and Galileo systems are operational with

24 satellites M and few satellites K, and at least 18 satellites adequately.

In this scenario for 2021 year all three systems GPS, GLONASS and Galileo are fully operational, all satellites of these systems transmit at least three frequencies accessible for civil users and the signals contain the integrity information. The GPS spatial segment will consist of at least 24 satellites of Block III, 12 satellites of Block IIF and perhaps all satellites of Block IIR–M and few of Block IIR. The spatial segments of GLONASS and Galileo will consist of 30 satellites of new Block K and 30 adequately.

In this optimistic scenario, already in 2016 year, all present-day SBAS, and GAGAN and QZSS also, will be fully operational, and perhaps in 2021 year other new systems (e.g. in Africa and in South America) additionally.

4.2 *Pessimistic scenario*

In this scenario no project is not realized according to earlier plan. In the case of GPS system the following assumptions are made for 2016 year:
- 8 Block IIF satellites were launched only,
- the block III did not begin,
- the vitality of all satellites on orbit are at most equal nominal. It means that the satellites of Block IIR and earlier are out of service.

In this situation we have in satellites: 8 Block IIFs ranging from 0 to 6 years old and 8 Block IIR–Ms ranging from 7 to 11 years old, that is to say 16 satellites only. It means that user's position cannot be obtained at any point on Earth and at any moment.

In 2016 year the number of Galileo satellites fully operational is less than planned 18, therefore this system is still under construction. The number of GLONASS satellites, all kind M, is less than 24 again. The works over the next satellite generation K continually last.

In scenario for 2021 year GPS spatial segment will consist at most of 12 satellites of Block IIF and few satellites of Block III only. As the vitality of all GPS satellites are at most equal nominal, the satellites of Block IIR–M and earlier are already out of service. The GLONASS spatial segment will consist at most of 24 satellites M and few satellites K only. The date of FOC (Full Operational Capability) of Galileo system continually lengthens, the number of satellites is less than nominal 27 still.

In this pessimistic scenario in 2016 and 2021 years EGNOS, WAAS and MSAS are fully operational, but without additional geostationary satellites. GAGAN and QZSS are under construction still.

4.3 *The most probable scenario*

All systems are undergoing modernization or construction, but time-limits are not kept.

The last launch of GPS IIF satellite and the first launch of GPS III satellite will be not in 2014 years, but several years later. The vitality of GPS satellites is continually the same as at present, for the most satellites greater than nominal. In this situation in 2016 year we have in satellites: at most 10 Block IIRs ranging from 12 to 16 years old, 8 Block IIR–Ms ranging 7 to 11 years old and 12 Block IIFs ranging from 0 to 6 years old. The construction of Galileo system became finished, but the number of satellites is 18 only. The spatial segment of the GLONASS system consists of 24 satellites M only.

In this scenario in 2021 year we have in GPS satellites: 12 Block IIIs ranging from 0 to 5 years old, 12 Block IIF ranging from 5 to 11 years old and about 8 Block IIR–Ms ranging from 12 to 16 years old. The Galileo system with the number of satellite between 27 and 30 is fully operational. The GLONASS system will consist of about 30 satellites M and K, in the most of the block M.

In the most probable scenario GAGAN and QZSS systems will be fully operational before 2016 year, but in 2021 year other new SBAS will be under construction or on the stage projects.

5 CONCLUSIONS

- in the case of GPS system the kind of scenario will depend on vitality of his satellites, of Block IIR–M in particular. If this vitality will be equal a dozen or so years, as in earlier blocks, scenario will be optimistic,
- in the case of the GLONASS and Galileo systems the kind of scenario will depend on time-limit of the implementation of all improvements,
- in optimistic scenario in 2021 GPS, GLONASS and Galileo systems offer full service on all 32, 24 and 27 satellites, adequately and information about integrity; five years earlier integrity provides the Galileo system only,
- in pessimistic scenario in 2021 one only SNS, the GLONASS system, offers full service, the number of GPS satellites is less than nominal 24, the Galileo system is under construction still; five years earlier all these three SNSs are not fully operational,
- in the most probable scenario in 2021 all three SNSs are fully operational, but in each system information about integrity can be obtained only from the part of his satellites; five years earlier this information is provided by the part of GPS and Galileo satellites only.

REFERENCES

Gleason S., Gebre-Egziabher D. 2009. GNSS Applications and Methods, Artech House, Boston/London.

Gower A. 2008. The System: The Promise of OCX, GPS World, No 8, vol.19.

Hofmann-Wellenhof B. et all. 2008. GNSS–Global Navigation Satellite Systems GPS, GLONASS, Galileo & more, SpringerWienNewYork, Wien.

Januszewski J. 2010. Visibility and geometry of combined constellations GPS with health in question, GLONASS and Galileo, p. 1082–1094, Institute of Navigation, International Technical Meeting, San Diego (CA).

Januszewski J. 2009. Satellite navigation systems integrity today and in the future, Monograph "Advances in Transport Systems Telematics", p. 123–132, Edited by Jerzy Mikulski, Wydawnictwa Komunikacji i Łączności, Warszawa.

Januszewski J. 2010. Nawigacyjny system satelitarny GPS dzisiaj i w przyszłości, p. 17–29, Prace Wydziału Nawigacyjnego nr 24, Akademia Morska, Gdynia (in polish).

Kaplan E.D. & Hegarty C.J. 2006. Understanding GPS Principles and Applications, Artech House, Boston/London.

Lavrakas J.W. 2007. A Glimpse into the Future: A Look at GNSS in the Year 2017, p. 210–217, Institute of Navigation, National Technical Meeting, San Diego (CA).

Roper E. 2010. GPS Status and Modernization, Munich Satellite Navigation Summit, Munich.

Springer T., Dach R. 2010. GPS, GLONASS, and More Multiple Constellation Processing in the International GNSS Service, GPS World, No 6, vol.21.

www.glonass-ianc.rsa.ru

www.navcen.uscg.gov

Table 1. GPS System, PRN – Pseudorandom noise number, SVN – Space Vehicle Number, launch and input dates, active life and mean active life in years and months of all 32 satellites in January 21, 2011

Block	PRN	SVN	Launch date	Input date	Active life		Mean active life	
					years	months	years	months
IIA–10	32	23	26.11.1990	10.12.1990	16	0.8		
IIA–11	24	24	04.07.1991	30.08.1991	19	1.3		
IIA–14	26	26	07.07.1992	23.07.1992	18	5.8		
IIA–15	27	27	09.09.1992	30.09.1992	18	3.2		
IIA–21	9	39	26.06.1993	20.07.1993	17	4.8		
IIA–23	4	34	26.10.1993	22.11.1993	17	2.0	16	3.1
IIA–24	6	36	10.03.1994	28.03.1994	16	9.0		
IIA–25	3	33	28.03.1996	09.04.1996	14	8.1		
IIA–26	10	40	16.07.1996	15.08.1996	14	4.4		
IIA–27	30	30	12.09.1996	01.10.1996	14	2.8		
IIA–28	08	38	06.11.1997	18.12.1997	13	0.2		
IIR–2	13	43	23.07.1997	31.01.1998	12	11.6		
IIR–3	11	46	07.10.1999	03.01.2000	11	0.6		
IIR–4	20	51	11.05.2000	01.06.2000	10	7.5		
IIR–5	28	44	16.07.2000	17.08.2000	10	5.2		
IIR–6	14	41	10.11.2000	10.12.2000	10	1.3		
IIR–7	18	54	30.01.2001	15.02.2001	9	11.1	8	11.0
IIR–8	16	56	29.01.2003	18.02.2003	7	10.9		
IIR–9	21	45	31.03.2003	12.04.2003	7	7.2		
IIR–10	22	47	21.12.2003	12.01.2004	7	0.3		
IIR–11	19	59	20.03.2004	05.04.2004	6	9.5		
IIR–12	23	60	23.06.2004	09.07.2004	6	6.3		
IIR–13	2	61	06.11.2004	22.11.2004	6	1.9		
IIR–14M	17	53	26.09.2005	13.11.2005	5	1.1		
IIR–15M	31	52	25.09.2006	13.10.2006	4	3.3		
IIR–16M	12	58	17.11.2006	13.12.2006	4	1.1		
IIR–17M	15	55	17.10.2007	31.10.2007	3	2.7	3	1.8
IIR–18M	29	57	20.12.2007	02.01.2008	3	0.6		
IIR–19M	7	48	15.03.2008	24.03.2008	2	9.9		
IIR–20M	1	49	24.03.2009	in commissioning phase				
IIR–21M	5	50	17.08.2009	27.08.2009	1	4.8		
IIF–1	25	62	28.05.2010	27.08.2010	0	4.8	0	4.8

Table 2 GLONASS System, orbit/slot, frequency channel, GLONASS number, launch and input dates, active life and mean active life in years and months of all 21 satellites in January 21, 2011

Orbit / slot	Frequency channel	GLONASS number	Launch date	Input date	Life time years	Life time months	Mean life time years	Mean life time months
I / 1	01	730	14.12.2009	30.01.2010	1	1.2		
I / 2	− 4	728	25.12.2008	20.01.2009	2	0.9		
I / 3		satellite 727 in maintenance						
I / 4		without satellite						
I / 5	01	734	14.12.2009	10.01.2010	1	1.2		
I / 6	− 4	733	14.12.2009	24.01.2010	1	1.2		
I / 7	05	712	26.12.2004	07.10.2005	6	0.9		
I / 8	06	729	25.12.2008	12.02.2009	2	0.9		
II / 9	− 2	736	02.09.2010	04.10.2010	0	4.6		
II / 10	− 7	717	25.12.2006	03.04.2007	4	0.9		
II / 11	00	723	25.12.2007	22.01.2008	3	0.9		
II / 12	− 1	737	02.09.2010	04.10.2010	0	4.6	2	2.2
II / 13	− 2	721	25.12.2007	08.02.2008	3	0.9		
II / 14 [x1]	− 7	722	25.12.2007	25.01.2008	3	0.9		
II / 15	00	716	25.12.2006	12.10.2007	4	0.9		
II / 16	− 1	738	02.09.2010	04.10.2010	0	4.6		
III / 17		satellites 714 and 728 in maintenance						
III / 18	−3	724	25.09.2008	26.10.2008	2	3.9		
III / 19	03	720	26.10.2007	25.11.2007	3	2.9		
III / 20	02	719	26.10.2007	27.11.2007	3	2.9		
III / 21	04	725	25.09.2008	05.11.2008	2	3.9		
III / 22 [x2]	-3	731	02.03.2010	28.03.2010	0	10.7		
III / 23	03	732	02.03.2010	28.03.2010	0	10.7		
III / 24	02	735	02.03.2010	28.03.2010	0	10.7		

[x1] – additional satellite 715 in maintenance, [x2] – additional satellite 726 in maintenance

Table 3. The most significant events in the satellite navigation systems and satellite based augmentation systems in the nearest 10 years and their consequences for users

Year	Event	Consequences for users
2010	three GLONASS M satellites crashed into Pacific Ocean after a failed launch	Full Operational Capability of GLONASS system cannot be obtained
	first launch of QZSS spacecraft Michibiki	for the first time in history the signal L1C is transmitted in space
	additional launches of Compass satellites	new GEO, IGSO and MEO satellites of China's system
2011	24 GLONASS M satellites	two SNS systems (GLONASS and GPS) fully operational
	all GAGAN satellites on geostationary orbit	GAGAN – Indian SBAS fully operational
	the first launch of GLONASS K satellite	the beginning of the new generation of GLONASS satellite
		the first use of code division multiple access CDMA
2012	third SDCM satellite on geostationary orbit	SDCM – Russian SBAS fully operational
2013	WAAS – two frequencies (L1 and L5) for ionospheric corrections	elimination of vertical guidance caused by ionospheric storms
2014	the first launch of GPS III A satellite	the beginning of the third generation of GPS system
2015	Full Operational Capability of the next Generation GPS Control Segment (OCX) Galileo constellation with 18 satellites (4 IOV and 14 fully operational)	continuous L-band tracking coverage of the GPS constellation additional features and functionality of control segment (CS) for the first time in history, integrity information about SNS for the users of the all the world, Initial Operational Capability (IOC)
2016	24 GPS satellites transmitting L2C	full access to two civil frequencies
2018	24 GPS satellites transmitting L5 Galileo constellation 27–30 satellites	full access to three civil frequencies full access to all signals and services, Full Operational Capability (FOC)
2019	30 GLONASS K satellites	full access to three civil frequencies, integrity information about system
2020	35 Compass satellites fully operational (5 GEO, 27 MEO and 3 IGSO)	full access to all signals and services
2021	24 GPS satellites block III transmitting L1C	full access to new block III, integrity information and new signal L1C

Table 4. The projected total number of satellites, number of satellites transmitting signals for civil users on two and three frequencies and information about integrity for different satellite navigation systems and for different scenarios in 2016 and 2021 years

Year	Scenario	System	Number of satellites			Integrity of the system
			total	with two frequencies	with three frequencies	
2016	optimistic	GPS	40	24	16	non
		GLONASS	at least 24	at least 24	several	non
		Galileo	at least 18	at least 18	at least 18	yes
		Total	at least 82	at least 66	at least 40	–
	pessimistic	GPS	16	8	8	non
		GLONASS	less than 24	less than 24	0	non
		Galileo	less than 18	less than 18	less than 18	non
		Total	less than 58	less than 50	less 26	–
	the most probable	GPS	at most 30	at most 20	12	non
		GLONASS	24	24	0	non
		Galileo	18	18	18	yes
		Total	at most 72	at most 62	30	–
2021	optimistic	GPS	at least 44	at least 44	24	yes
		GLONASS	30	30	30	yes
		Galileo	30	30	30	yes
		Total	at least 104	at least 104	84	–
	pessimistic	GPS	a dozen or so	a dozen or so	a dozen or so	non
		GLONASS	at most twenty several	at most 24	several	non
		Galileo	less than 27	less than 27	less than 27	non
		Total	about 65	about 60	about 45	–
	the most probable	GPS	about 32	about 32	24	non
		GLONASS	about 30	a dozen or so	a dozen or so	non
		Galileo	27 ÷ 30	27 ÷ 30	27 ÷ 30	yes
		Total	about 89 ÷ 92	about 73 ÷ 79	about 63 ÷ 72	–

2. GNSS Meteorology

J. Bosy, W. Rohm, J. Sierny & J. Kaplon
Wroclaw University of Environmental and Life Sciences

ABSTRACT: GNSS meteorology is the remote sensing of the atmosphere (troposphere) using Global Navigation Satellite Systems (GNSS) to derive information about its state. The most interesting information is a delay of the signal propagation due to the water vapor content - the Slant Wet Delay (SWD). The inverse modeling technique being concern here is the tomography. It is the transformation of the slant integrated observation of state of the atmosphere (SWD), to the three dimensional distribution of the water vapor. Over past six years the studies on GNSS tomography were performed in the Wroclaw University of Environmental and Life Sciences on the GNSS tomography. Since 2008 the new national permanent GNSS network ASG-EUPOS (about 130 GNSS reference stations) has been established in Poland (www.asgeupos.pl). This paper presents the issues of the Near Real Time troposphere model construction, characteristic of GNSS and meteorological data and the building of the required IT infrastructure.

1 INTRODUCTION

Global Navigation Satellite System is designed for positioning, navigation, amongst other possible applications it can also be used to derive information about the state of the atmosphere, what is now recognized as GNSS meteorology. Particularly GNSS meteorology is the remote sensing of the atmosphere from satellite platform (GNSS radio occultation meteorology) (Pavelyev et al. 2010) and ground permanent stations (ground based GNSS meteorology) (Bender et al. 2010). Continuous observations from GNSS receivers provide an excellent tool for studying the earth atmosphere. There are many GNSS meteorology applications: climatology, nowcasting and 4D monitoring.

The ground based GNSS meteorology is based on the tropospheric delay, one of the results of GNSS data processing . The tropospheric delay is represented by the Zenith Total Delay ZTD. The ZTD can be split into hydrostatic ZHD and wet ZWD component of the delay:

$$ZTD = ZHD + ZWD \qquad (1)$$

The wet component of Zenith Tropospheric Delay ZWD is the foundation for computing of water vapor content in the atmosphere. The relation between ZWD and the water vapor content in atmosphere is expressed by IWV (Integrated Water Vapor) and given by the equation (Kleijer 2004):

$$IWV = \frac{ZWD}{10^{-6} \cdot R_w} \left(k'_2 + \frac{k_3}{T_M} \right)^{-1} \qquad (2)$$

where R_w is the specific gas constant for water vapor, k'_2, k_3 are refraction constants (Boudouris 1963) and T_M is weighted mean water vapor temperature of the atmosphere (Kleijer 2004).

The IPWV (Integrated Precipitable Water Vapor) is computed IPWV according to relation:

$$IPWV = \frac{IWV}{\rho_w} \qquad (3)$$

where ρ_w is the water density (Mendes 1999).

The *IPWV* is delivered according to equations (2 and 3) from *ZWD* and gives the information about contents of water vapor (2D model) above GNSS stations. The EUREF Permanent Network (EPN: www.epncb.oma.be) is the base of determination of IPWV in Europe (Vedel and Huang 2004). Since 2005 EPN analysis centres ASI, BKG, GOP and LPT delivers Near Real Time *ZTD* for meteorological applications in the frame of international project E-GVAP (EUMETNET GPS Water Vapour Programme) (Dousa 2010).

The spatial structure and temporal behavior of the water vapor in the troposphere (4D model) can be modeled by using the GNSS tomography method. The input data of GNSS tomography are: the signal Slant Wet Delays SWD, which are the results of the

GNSS data processing, the meteorological observations from synoptic stations and the Numerical Weather Prediction (NWP) models data. The NWP models data are also used for GNSS data verification and calibration of the tomography model (Rohm and Bosy 2010). The STD can be separated like (1) into hydrostatic SHD and wet SWD components and represented by the well known relation:

$$STD = SHD + SWD = m_d(\varepsilon)ZHD + m_w(\varepsilon)ZWD \quad (4)$$

where ε is the satellite elevation angle and $m_d(\varepsilon)$ and $m_w(\varepsilon)$ are the mapping functions (Niell 1996; Boehm et al. 2006).

In the GNSS tomography SWD extracted from (4) is linked with the wet refractivity N_w by the given equation:

$$SWD = A \cdot N_w \quad (5)$$

where A is the design matrix.

Currently several methods exist to solve the GNSS tomography model. The first is to add horizontal and vertical constraints into the system of equations (5) and then solve it (Hirahara 2000), the second is to use a Kalman filter with the same equation system (Flores et al. 2000), the third is to find the solution directly from the GNSS phase measurement equation (Nilsson and Gradinarsky 2006) and another is Algebraic Reconstruction Technique (ART) developed by Kaczmarz (Bender et al. 2009). The method presented in this paper uses the minimum constraint conditions imposed on the system of observation equations (5) (Rohm and Bosy 2009; Rohm and Bosy 2010).

The wet refractivity N_w is estimated from equation (5) and finally the water vapour distribution in the troposphere (4D) represented by the water vapour partial pressure e and the temperature T is extracted from the formula:

$$N_w = \left(k_2' \frac{e}{T} + k_3 \frac{e}{T^2} \right) Z_v^{-1} \quad (6)$$

where Z_v^{-1} is an inverse empirical compressibility factor (Owens 1967).

The new Polish national permanent GNSS network (Ground Base Augmentation System) ASG-EUPOS has been established since 2008. 17 Polish stations equipped with GNSS receivers and uniform meteorological sensors work currently in the frame of the European Permanent Network (Bosy et al. 2007; Bosy et al. 2008). The ASG-EUPOS network consists (including foreign stations) of about 130 GNSS reference stations located evenly on the country area and build network of greater density than EPN network. This guarantees that the 4D troposphere delay and water vapor models will be more representative for the territory of Poland.

Since 2010 the idea of integrated researches based on the GNSS and meteorological observations

from ASG-EUPOS stations is realized in the frame of research project entitled *Near Real Time atmosphere model based on the GNSS and the meteorological data from the ASG-EUPOS reference stations on the territory of Poland*. The paper presents in the second section the methodology of NRT atmosphere models construction procedures. The second section encloses proposal of the method of water vapor distribution in space and time (4DWVD) using GNSS tomography technique. The third section includes the ASGEUPOS system description and sources of GNSS and meteorological data, localization and accuracies. The fourth section contains the specification of IT infrastructure for NRT data streaming and processing. The paper is closed in fifth section with conclusions.

2 NEAR REAL TIME ATMOSPHERE MODEL

The GNSS and meteorological observations form ASG-EUPOS stations are the base of near real time models of tropospheric delay and water vapor (NRT ZTD and NRT ZWD) in atmosphere. Figure 1 shows the diagram of NRT ZTD and NRT ZWD models construction (Bosy et al. 2010).

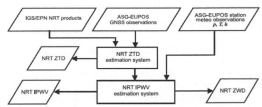

Figure 1: The diagram of NRT ZTD, ZWD and IPWV models construction on the base of GNSS and meteorological data from ASG-EUPOS reference stations

The NRT ZTD will be obtained from the NRT solution of ASG-EUPOS stations network. The strategy of NRT solution will be realized according to standards used for global IGS and regional EPN permanent GNSS networks and NRT solution strategy created in the frame of COST Action 716 (European Cooperation in the field of Scientific Technical Research-exploitation of ground-based GPS for climate and numerical weather prediction applications, 1998-2004), TOUGH (Targeting Optimal Use of GPS Humidity Data in Meteorology, *http://tough.dmi.dk/*, 2003-2006) and E-GVAP (The EUMETNET GPS Water Vapour Programme, *http://egvap.dmi.dk*, 2004-2008) projects (Dousa 2004; Dousa 2010). The ZHD for all ASG-EUPOS stations will be estimated in NRT mode on the base of meteorological observation of Polish EPN stations equipped with meteorological sensors. Next according to relation (1) the values of ZWD will be computed. The IWV and IPWV values above all

ASG-EUPOS stations will be calculated from equations (2) and (3) and finally NRT ZWD and NRT IPWV models for Poland territory will be constructed (Bosy et al. 2010).

The spatial structure and temporal behavior of the water vapour in the troposphere (4D) can be modeled using the GNSS tomography method. The GNSS signal delays due to the water vapour are evaluated for a large number of different views through the atmosphere (Bender and Raabe 2007). The idea of GNSS tomography for Poland is presented in the figure 2 (Bosy et al. 2010).

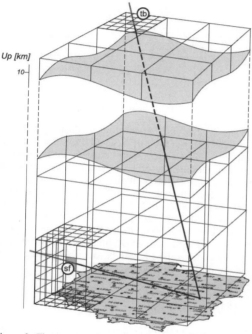

Figure 2: The ray path in consecutive voxels. Two cases are considered, the first when the ray is coming out of the model's side face (sf), and the second, when ray is comming out of the model top boundary (tb)

The input data of GNSS tomography are: the signal Slant Wet Delays *SWD*, which are the results of the GNSS data processing, the meteorological observations from ASG-EUPOS meteo stations. The quality and quantity of the GNSS observations is strongly correlated with the GNSS satellites constellation, the number of the ground stations and inter-station distances. As a result of the GNSS data processing the cut off angle of *SWD* observations are set to 3 – 5 degrees of elevation. Moreover, the empirical results show that the angles are ranging from 5 to 85 degrees, but most of the observations are clustered between 10 and 15 degrees. In case of GNSS troposphere tomography it is typical condition

(Bender and Raabe 2007). In the GNSS tomography *SWD* is linked with wet refractivity N_w by the equation (5). One of the method to resolution of equation (5) is the authors method (Rohm and Bosy 2009; Rohm and Bosy 2010) presented briefly below.

To find the voxels' (Fig. 2) refractivities one needs to invert the equation (5), which in theory might be solved by the means of the least squares method:

$$N_w = \left(A^T \cdot P \cdot A\right)^{-1} \cdot A^T \cdot P \cdot SWD^T \qquad (7)$$

where the A is design matrix (5) and P is a weighting matrix. The weighting matrix P is constructed as an inversion of covarince matrix of observations *SWD* given by the formula:

$$P = C_{SWD}^{-1} \qquad (8)$$

To get 3D picture of the wet refractivity N_w the Singular Value Decomposition (*SVD*) technique is used. The *SVD* technique is the pseudo inverting of system (5) on the base of factorization of the variancecovariance matrix in the equation (7).

$$\left(A^T \cdot P \cdot A\right)^+ = V \cdot S^+ \cdot U^T \qquad (9)$$

where U is a $n \times n$ orthogonal matrix of left-singular vectors, V is a $m \times m$ orthogonal matrix of right singular vectors, S is a $n \times m$ diagonal matrix of singular values sorted in descending order (Anderson et al. 1999) and S^+ is a pseudoinverse of the matrix S.

3 ASG-EUPOS GNSS AND METEOROLOGICAL DATA

The GNSS data are currently available from the GNSS permanent stations operated in the frame of national networks. In Poland the national permanent GNSS network ASG-EUPOS has been established since 2008. The receiving segment (ground control segment) consists of a network of GNSS reference stations located evenly on the whole territory of Poland. Comply with EUPOS and project of the ASG-EUPOS system standards distances between neighboring reference stations should be 70 km what gives number of stations 98 (3). According to rules of EUPOS organization (in the frame of cross-border data exchange) 3 reference stations from Lithuania (LITPOS), 6 stations from Germany (SAPOS), 7 stations from Czech Republic (CZEPOS) and 6 stations from Slovakia (SKPOS) were added (Fig. 3) (Bosy et al. 2008).

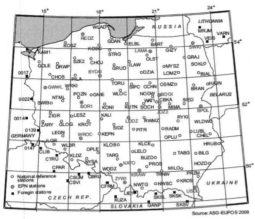

Figure 3: Reference stations included of ASG-EUPOS system (www.asgeupos.pl)

The reference stations of ASG-EUPOS system (Fig. 3) are equipped with the modern GNSS receivers and the antennas with absolute calibrations. In the 14 localizations of the EPN stations the new uniform meteorological infrastructure Paroscientic, Inc. MET4A sensors were installed. In the EPN/IGS station Borowiec (BOR1) the equivalent meteorological sensors: NAVI Ltd. HPTL.3A and Skye Instruments Ltd. are installed.

Meteorological observations from ASG-EUPOS stations (Bosy et al. 2010) will be used for ZHD extraction from equation (1), to compute ZWD and finally SWD (4). The external meteorological observations from Institute of Meteorology and Water Management (IMGW) synoptic stations, radiosoundings observations and NWP COAMPS model outputs will be used also for verification of GNSS tomography model (Rohm and Bosy 2010).

4 IT INFRASTRUCTURE

Meteorological data from stations dispersed on the area of study are collected with GNSS data and put into ASG-EUPOS caster. Then using the NTRIP (Networked Transport of RTCM via Internet Protocol), data are transferred to the center that deals with NRT (Near Real Time) processing. The Internet is particularly well suited to transmit data between different providers over long distances. However, the servers required must be tied to the Internet via interconnected broadcasters with sufficient bandwidth. In order to secure resources the infrastructure must be equipped with firewall and IDS (Intrusion Detection System) / IPS (Intrusion Prevention System) elements (Fig. 4).

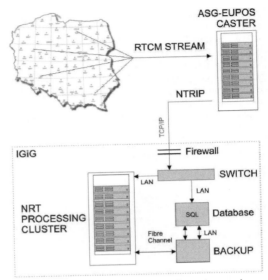

Figure 4: IT infrastructure for data streaming and processing

The GNSS and meteorological data from Management Centre (MC) of ASG-EUPOS are decoded and sent both to compute clusters and database (Fig. 4). Despite the effectiveness of the algorithms computations require compute clusters. That is why one needs such computer cluster, which enables parallel computations. NRT processing should be equipped with back-up system, which connects with other devices through dedicated high-speed fiber channel network. The above presented conception is still very general and will be detailed during tests and the project realization.

5 CONCLUSIONS

Ground based GNSS meteorology currently utilizes the height resolution GBAS networks like AS-GEUPOS, where reference stations are equipped with GNSS and meteorological sensors. The NRT troposphere model based on the GNSS and meteorological data of GBAS system could be used as well as in meteorological applications, in the real-time and post-processing positioning services of AS-GEUPOS system. The model created from meteorological and GNSS data, could be competitive to Numerical Weather Prediction models, especially for nowcasting. The improvement in positioning is that tropospheric delays will be calculated directly from observations, not like now from deterministic models.

ACKNOWLEDGEMENTS

This work has been supported by the Polish Ministry of Science and Higher Education: research project No N N526 197238.

REFERENCES

Anderson, E., Z. Bai, C. Bischof, S. Blackford, J. Demmel, J. Dongarra, J. Du Croz, A. Greenbaum, S. Hammarling, A. McKenney, and D. Sorensen (1999). *LAPACK Users' Guide, Third Edition.* Society for Industrial and Applied Mathematics.

Bender, M., G. Dick, J. Wickert, M. Ramatschi, M. Ge, G. Gendt, M. Rothacher, A. Raabe, and G. Tetzlaff (2009). Estimates of the information provided by GPS slant data observed in Germany regarding tomographic applications. *J. Geophys.* Res. 114, D06303.

Bender, M. and A. Raabe (2007). Preconditions to ground based GPS water vapour tomography. *Annales Geophysicae* 25(8), 1727–1734.

Bender, M., R. Stosius, F. Zus, G. Dick, J. Wickert, and A. Raabe (2010). Gnss water vapour tomography – expected improvements by combining gps, glonass and galileo observations. *Advances in Space Research In Press, Corrected Proof, –.*

Boehm, J., A. Niell, P. Tregoning, and H. Schuh (2006). Global Mapping Function (GMF): A new empirical mapping function based on numerical weather model data. *Geophys. Res. Lett. 33*, 15–26.

Bosy, J., W. Graszka, and M. Leonczyk (2007). ASGEUPOS. A Multifunctional Precise Satellite Positioning System in Poland. *European Journal of Navigation 5(4)*, 2–6.

Bosy, J., A. Oruba, W. Graszka, M. Leonczyk, and M. Ryczywolski (2008). ASG-EUPOS densification of EUREF Permanent Network on the territory of Poland. *Reports on Geodesy 2(85)*, 105–112.

Bosy, J., W. Rohm, and J. Sierny (2010). The concept of the near real time atmosphere model based on the GNSS and the meteorological data from the ASG-EUPOS reference stations. *Acta Geodyn. Geomater.* 7(1(157)), 1–9.

Boudouris, G. (1963). On the index of refraction of air, the absorption and dispersion of centimeter waves by gases. *Journal of Research of the National Bureau of Standards 67D(6)*, 631684.

Dousa, J. (2004). Evaluation of tropospheric parameters estimated in various routine GPS analysis. *Physics and Chemistry of the Earth, Parts A/B/C 29*(2-3), 167 – 175. Probing the Atmosphere with Geodetic Techniques.

Dousa, J. (2010). The impact of errors in predicted GPS orbits on zenith troposphere delay estimation. *GPS Solutions.*

Flores, A., G. Ruffini, and A. Rius (2000). 4D tropospheric tomography using GPS slant wet delays. *Annales Geophysicae 18(2)*, 223–234.

Hirahara, K. (2000). Local GPS tropospheric tomography. *Earth Planets Space 52*(11), 935–939.

Kleijer, F. (2004). *Troposphere Modeling and Filtering for Precise GPS Leveling.* Ph. D. thesis, Department of Mathematical Geodesy and Positioning, Delft University of Technology, Kluyverweg 1, P.O. Box 5058, 2600 GB DELFT, the Netherlands. 260 pp.

Mendes, V. B. (1999). *Modeling the neutral-atmosphere propagation delay in radiometric space techniques.* Ph. D. thesis, Deparment of Geodesy and Geomatics Engineering Technical Reort No. 199, University of New Brunswick, Fredericton, New Brunswick, Canada.

Niell, A. E. (1996). Global mapping functions for the atmosphere delay at radio wavelenghs. *J. Geophys. Res. 101*(B2), 3227–3246.

Nilsson, T. and L. Gradinarsky (2006). Water Vapor Tomography Using GPS Phase Observations: Simulation Results. *IEEE Trans. Geosci. Remote Sens. 44*(10 Part 2), 2927–2941.

Owens, J. (1967). Optical refractive index of air: dependence on pressure, temperature and composition. *Appl. Opt. 6*(1), 51–59.

Pavelyev, A., Y. Liou, J.Wickert, T. Schmidt, and A. Pavelyev (2010). Phase acceleration: a new important parameter in gps occultation technology. *GPS Solutions 14*, 3–11. 10.1007/s10291-009-0128-1.

Rohm, W. and J. Bosy (2009). Local tomography troposphere model over mountains area. *Atmospheric Research 93*(4), 777 – 783.

Rohm, W. and J. Bosy (2010). The verification of gnss tropospheric tomography model in a mountainous area. *Advances in Space Research In Press, Corrected Proof, –.*

Vedel, H. and X. Huang (2004). Impact of ground based GPS data on Numerical Weather Prediction. *J. Meteor. Soc. Japan 82*(1B), 459–472.

3. Onboard Wave Sensing with Velocity Information GPS

Y. Arai
Marine Technical College, JAPAN

E. Pedersen
Norwgian University of Science and Technology, NORWAY

N. Kouguchi
Kobe University, JAPAN

K. Yamada
ex Hitachi Zosen Corporation, JAPAN

ABSTRACT: Even though it is essential that the wave information promotes greater safety also efficiency to navigate and/or operate a ship not only ocean going but also docking/landing, it is very difficult to sense a wave information such as wave heights and periods or wave lengths in real time at the present time.   On the other hand, the Velocity Information (VI) GPS is developed as the stand-alone 3D velocities measurement equipment of which accuracies are precise (less than 1 cm/sec.) and the coverage is all over the world.   It is able to drive Wave information from not only the time history of wave amplitudes but also the time history of wave velocities.   The algorism to sense the wave information such as not only significant wave height but also plural wave heights and period intervals of encounter was presented by the authors in IAIN2009 and ANC2010.   In this paper, the introduction and the performance of wave sensing with VI GPS and some results of onboard experiments are described and discussed.

1 INTRODUCTION

SDME (Speed and Distance Measurement Equipment), which presents the two axes velocities OG (Over the Ground) to ship master and/or pilot, such as Doppler SONAR, etc. is manufactured as the best application of safety docking to dolphin or berth for VLCC, etc.

Recently it is also developing to assist maneuvering in approaching to dolphin or berth using high accuracy DGPS or RTK-GPS which requires communication to base-station, etc., and now the new technology on GPS, which calls VI-GPS (Velocity Information GPS) and presents very high accuracy velocities stand-alone or without communication to base-station, is coming to onboard application (Tatsumi, et al., 2009), (Yoo, et al., 2009), and Okuda, et al. (2008) presented the trade-off between accuracy and response in application of docking velocity that means the performance of Velocity Information is affected not only by the accuracy but also by the relationship between the time lag of SDME and her math or time constant of maneuverability.

Meanwhile, recently it is often taken to dock not only to berth but also to navigating vessel which calls STS (Ship To Ship) operation (Yoo, et al., 2009).   In case of STS operation, it is taken on open-sea or deep-sea, so the disturbances such as current, wave and/or wind affect ship maneuvering, and onboard sensing current, wave effect and wind are essential to make a good solution not only for safety but also for efficiency, or it is the trade-off between safety and economical issues.   Although it is very important issues to efficiently apply the SDME, but onboard sensing current and/or wave effect is very difficult because of the low responsibility and/or poor performance of STW (Speed Through the Water) (Arai, et al. 1983).   So, Arai, et al. (2009), (2010) developed the algorism to sense the current and wave effect without two axes SDME TW (Through the Water).

In this paper, at first the introduction of VI GPS, at second section the developed algorism to sense external forces or disturbances such as wave and current is introduced, at third section the onboard experiments, results and evaluations are presented, and finally it is concluded that proposed algorism and availability of onboard wave sensing with VI-GPS will be essential to ship operation not only docking or STS operation but also ocean going, etc.

2 VELOCITY INFORMATION GPS

2.1 *Outline of VI-GPS*

VI-GPS consists of front-end GPS receiver which is able to measure carrier phases at every epoch and processing unit which has the differential system of carrier phases and phased and/or coded GPS positioning system.

VI-GPS positioning system is same as conventional system, so it is able to position fix within the accuracy of several meters. In the conventional GPS the measurement of carrier frequency with Doppler shift drives the velocities or SOG (Speed Over the Ground) and COG (Course Over the Ground), but VI-GPS measures carrier phase at every epoch Φ_i (m) and calculates the time difference of carrier phases between serial epochs $\delta\Phi_i$ is following (Tatsumi, et al. 2008):

$$\delta\Phi_i = \Phi_i - \Phi_{i-1}$$
$$= \delta\rho + c \cdot (\delta dt - \delta dT) + \epsilon_{\delta\Phi} \tag{1}$$

where, ρ is the geometric distance between a satellite and a receiver (m); c is the light speed in vacuum (m/s); dt and dT are the receiver and satellite clock error (s); $\epsilon_{\delta\Phi}$ is the measurement noise and errors which are not able to be modelled; and the symbol δ is the time difference operator.

Time differential observation drives cancelling propagation errors and little clock errors in the receiver and satellite, so VI-GPS is a stand-alone system which presents high accuracies of velocities without referential stations, etc. all over the world

2.2 Accuracies of Ship's Velocities

The essential maneuvering information is categorized and shown in Figure 1. In this figure, Heading, ROT (Rate Of Turning) and Wind Speed and Direction are measured by the conventional instruments, and controllable parameters of ship maneuvering are rudder motion (RUD) and/or propeller revolution (RPM), etc. Ship's speed SOG and course COG are measured by two axes SDME OG such as Doppler SONAR, etc.

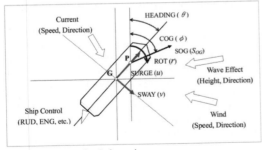

Figure 1. Maneuvering Information.

To maneuver for safety and economically especially in docking, longitudinal and lateral velocities at a Setting Point P_S, and the disturbances such as wind, current and/or wave information are essential. The relationship between velocities at Setting Point $P_S = (X_S, Y_S)$ and Surge/Sway at the Center of Ship O are following (Arai, et Al. 2010):

$$U_{lon} = u - r \cdot Y_S$$
$$U_{lat} = v + r \cdot X_S \tag{2}$$

where, U_{lon} and U_{lat} are longitudinal and lateral velocities at Setting Point of the sensor P_S; $U = (u, v)$ u and v are surge and sway (m/s); and r is Rate of Turn (rad./s).

In case of using VI-GPS, it is able to measure SOG/COG, so the relationship between SOG/COG at Sensor Position $P_{GPS} = (X_{GPS}, Y_{GPS})$ is following:

$$u_{OG} = S_{OG} \cdot \cos(\emptyset - \theta) + r \cdot Y_{GPS}$$
$$v_{OG} = S_{OG} \cdot \sin(\emptyset - \theta) - r \cdot X_{GPS} \tag{3}$$

where, $U_{OG} = (u_{OG}, v_{OG})$ is the vector of two axes velocities OG at center of ship O.

Considering the error propagation from sensor error, the deviations of surge and sway (du_{OG}, dv_{OG}) are resolved and following with total differential equation:

$$du_{OG} = du_{GPS} + v \cdot (d\theta + r \cdot \Delta t) + Y_{GPS} dr \cdot (1 + \Delta t)$$
$$dv_{OG} = dv_{GPS} - u \cdot (d\theta + r \cdot \Delta t) - X_{GPS} dr \cdot (1 + \Delta t) \tag{4}$$

where, deviation of longitudinal/lateral velocity by VI-GPS is $(du_{GPS} = dV_{NS}\cos\theta + dV_{EW}\sin\theta)$ and ($dv_{GPS} = dV_{EW}\cos\theta - dV_{NS}\sin\theta$); V_{NS} and V_{EW} are the velocities of N-S and E-W direction components by GPS; and Δt is time difference between VI-GPS and Compass.

The variances of surge and sway are driven by Equation 4 and following, if every parameter is independent statistically:

$$\sigma_{u_{OG}}{}^2 = \sigma_{GPS}{}^2 + v^2\sigma_\theta{}^2 + Y_{GPS}{}^2\sigma_r{}^2$$
$$\sigma_{v_{OG}}{}^2 = \sigma_{GPS}{}^2 + u^2\sigma_\theta{}^2 + X_{GPS}{}^2\sigma_r{}^2 \tag{5}$$

where, the variances of VI-GPS, Heading and ROT are $\sigma_{GPS}{}^2$, $\sigma_\theta{}^2$ and $\sigma_r{}^2$.

According to Equation 4 and 5, the countermeasure to decrease the deviations and variances of surge and sway are described as follows:

1 To install GPS antenna at same point as setting point where ship master requires to docking.

2 In case of STS operation longitudinal velocity is not so slow, so it affects the accuracy of lateral velocity with the performance of compass.

3 Synchronizing between sampling time of VI-GPS and heading/ROT information is essential to keep high accuracy in ship's turning.

4 In case of sensing wave effect, to install GPS antenna at the ship's center.

SOG or STW (Speed Through the Water) according to logical consideration such as sensor position and setting point as discussed Equation 2 to 5, but also it is essential to consider fitting error of sensor, effect of wake in STW, etc. (Arai, et al., 1983).

2.3 Performance of VI-GPS

Two axes velocities were measured by VI-GPS, and almost cases are seemed as good, but in the stage of converting to surge and sway from COG and SOG (Equation 3) there are some troubles which caused by mismatching of timing between onboard communication systems in Compass and VI-GPS. This problem should be surveyed using correlation function or another method, and the time difference or synchronization should be within minimum effects. So, in case of high ROT such as under turning, the effect of mismatching would increase and it would be difficult to maintain high accuracy.

The performance of VI-GPS is following and numerical performance is shown in Table 1:
1. VI-GPS works stand-alone and presents high accuracy velocities or SOG and COG.
2. VI-GPS has a good response to measure as shown in Table 1, but total system response should be limited because of on-board Navigational Information system's data interval.
3. Accuracy of VI-GPS is excellent, and onboard measuring two axes velocities essential with heading information such as Gyrocompass is available.
4. RAIM (Radio Autonomous Integrity Monitoring) function is much important to gain the reliability of ship maneuvering.

Table 1. Performance of VI-GPS

The accuracy of Velocity	less than 1cm/s
Sampling Time	5 Hz (0.2 s)
Responsiveness	less than 1 s
Stand alone	Yes
Coverage	All over the world

3 ALGOLISM FOR ONBORD WAVE SENSING

3.1 Effect of the External Forces

The external forces which affect ship operation are wind, current and wave. The effect of wind is easily able to be sensed in real time and to be countermeasured for safety navigation, but it is very difficult to sense the effects of current and/or wave in real time. The effects of current and wave $U = (u, v)$ are included in the difference between surge/sway OG $U_{OG} = (u_{OG}, v_{OG})$ which are easily measured in high accuracy using two axes SDME OG such as Doppler SONAR and VI-GPS, and surge/sway TW $U_{TW} = (u_{TW}, v_{TW})$ which are not so easy able to be measured.

Current is steady for short term, so the current components (speed and/or direction) will be easily able to be sensed using LPF (Low Pass Filter) or moving average method. After sensing current, the wave effect component will be sensed to subtract from current component.

The advanced algorism shown as Figure 4 which has been developed to improve the demerit of former algorism which is be able to sense the wave direction not all around but the limit measurement only for 90 degrees and the average wave length. So we developed the advanced algorism which is able to sense the direction, length and height of plural waves using the Fourier Transform or FFT. The disturbance forces except DC or very low frequency components or current are following:

$$u = - \sum_i F_{lon}(\omega_i) \cdot A_{lon_i} \cdot \omega_i \cdot \sin(-\omega_i t + \varphi_i)$$

$$v = - \sum_i F_{lat}(\omega_i) \cdot A_{lat_i} \cdot \omega_i \cdot \sin(-\omega_i t + \varphi_i)$$

$$\omega_i = \omega_{o_i} - K_i \cdot S_{OG} \cdot \cos(\beta_i - \emptyset) \qquad (6)$$

where, angular frequency of encounter is ω_i (rad./s); angular frequency of wave is ω_{o_i} (rad./s); period of encounter is $T_i = 2\pi/\omega_i$ (s); wave length of encounter $\lambda_i = 2\pi/K_i$ (m); wave direction β_i (deg.); wave amplitude of longitudinal/ lateral components are A_{lon_i}, A_{lat_i} (m); response function of longitudinal/lateral ship motion are $F_{lon/lat}(\omega_i)$ and i means i-th wave components.

Surge/sway are affected by ship's motion, and it is possible to solve the response functions or models of longitudinal and lateral motions of ship using differential equations. It will be assumed that these response models are primary response, so response functions are defined as follows:

$$F_{lon/lat}(\omega_i) = 1/\sqrt{1 + (T_{lon/lat} \cdot \omega_i/2\pi)^2} \qquad (7)$$

where, $T_{lon/lat}$ is time constant for each response model.

The values $U = (u, v)$ are observed during proper time, then the wave components which mean spectrum amplitudes $V_{lon/lat}(\omega_i)$ and phases $\psi_{lon \atop lat}(\omega_i)$ are resolved by FFT. So, it is able to drive the wave information from these data and they are following:

$$V(\omega_i) = \sqrt{V_{lon}(\omega_i)^2 + V_{lat}(\omega_i)^2}$$

$$B(\omega_i) = \text{Atan2}[\text{sign}_{lon} \cdot V_{lon}(\omega_i), \text{sign}_{lat} \cdot V_{lat}(\omega_i)] \qquad (8)$$

where, $V(\omega_i)$ and $B(\omega_i)$ are speed and direction of water particles by wave at each frequency; and $\text{sign}_{lon/lat}$ means sign of $V_{lon/lat}(\omega_i)$ according to $B(\omega_i)$.

The examples of FFT results are following:

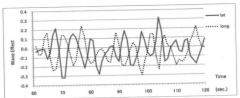

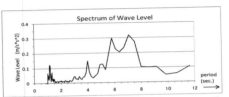

Figure 2. The time series of wave effect or U are sampled at every one second during one minute. Plural waves are found and roughly reading shows that the average period is less than 10 sec.

Spectrum of Wave Level

Figure 3. Sample Result of FFT (128 points). The unit of amplitude spectrum is shown as "m/s*s^2", it means amplitude of velocity at every $\sqrt{Hz}$. The unit of amplitude spectrum should be "(m/s)/$\sqrt{Hz}$". The result clearly shows that five waves exist and their periods. The longest one is 7.5 sec. and the shortest one is 1.5 sec.

3.2 Algorism of Onboard Wave Sensing

The flow chart of proposed algorism is shown in Figure .4, and consists of 5 parts as follows.

1 MMG: this part calculates to gain the two axes velocities TW U_{TW} which are longitudinal velocity u_w (here we call as Surge TW) and lateral velocity v_w (Sway TW) without effect of current and wave or only calculated from propeller thrust (ENG), rudder motion (RUD) and wind effect (Arai, et al., 1990).

2 CONVERT: Measured COG ϕ and SOG S_{OG} are converted using heading θ and sensor position to gain the two axes velocities OG U_{OG} which are longitudinal velocity OG u_{OG} (Surge OG) and lateral velocity OG v_{OG} (Sway OG).

3 SUB: To gain the current including wave effect, the subtraction from U_{OG} to U_{TW} and is executed. $U = U_{OG} - U_{TW}$: U consists of the velocity components of external forces which are affected ship's math, and is named as disturbances. Af-

ter subtraction, the coordinate system of U is converted from the ship coordinates to the terrestrial ones.

4 SPLITTER: It is done to split from U to current Uc and wave effect Uw using the Low Pass Filter (LPF) and the High Pass Filter (HPF), etc. Output from LPF consists of the current component Uc and cutoff frequency of LPF should be set as lower as possible and usually be set as 1/30 or 1/60 Hz which is lower than wave frequency.

5 FFT: FFT (Fast Fourie Transform) gains the frequency component of wave effect Uw. From the results of FFT, the amplitude and direction of wave effect Uw with respect to each frequency or period of wave are possible to be sensed according to Equation 6.

3.1 Accuracies of Wave Information

The wave amplitude at each frequency is drawn from the wave velocities Equation 6~8 and shown as Equation 9.

$$A(\omega_i) = \frac{U(\omega_i)}{F(\omega_i) \cdot \omega_i} \qquad (9)$$

So, it is able to drive the accuracy of wave amplitude which is following.

$$dA = \frac{dU}{F(\omega_i) \cdot \omega_i} - \frac{A(\omega_i) \cdot d\omega}{\omega_i} \qquad (10)$$

$$\sigma_{A_i} = \frac{1}{\omega_i} \sqrt{\frac{\sigma_U^2}{F(\omega_i)^2} + A(\omega_i)^2 \cdot \sigma_\omega^2} \qquad (11)$$

$$\sigma_\omega = 2\pi \cdot f_S / N \qquad (12)$$

where, σ_A, σ_U and σ_ω are root mean square of variances of wave amplitude, measured velocity and frequency; f_S is sampling frequency (Hz); and N is sampling number of FFT.

According to Equation 11, the variance of sensed wave height consists of two parts. The former is affected by the ship's mass or the response of ship's motion, and the latter is proportional one as sampling parameters.

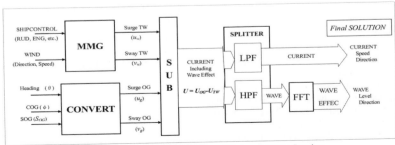

Figure 4. Flow Chart on Algorism of Onboard Wave Sensing

Using FFT, the every span of sequential frequencies is same. In case of short period, the number of wave in the width of periods increases. So, it is expected that the accuracy of wave information will increase according to averaging in the width of periods, and these are following:

$$\sigma_{A_j}{}' = \sigma_{A_j}/\sqrt{INT(4/(4j^2-1) \cdot N\Delta t/\Delta T}$$ (13)

where, $\Delta t = 1/f_S$ (seconds); ΔT is period interval (seconds); and INT means integer of ( ).

1 In case of onboard experiment, accuracy of wave height is within approximately 10 cm until period is under 15 seconds and it is available to observe the wave information.
2 Using VI-GPS, it is not necessary to make a data-link to base station such as RTK-GPS.
3 In case of large ship such as VLCC, the accuracy is not so good (1.5 meters) because of her ship's response.

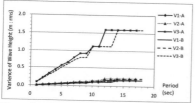

Figure 5. Expected variances of Wave Height. (V1: 150GT T/S, V2:500GT T/S, V3:VLCC) and (A means "f_S = 5 Hz, N=1024", B means "f_S=1 Hz, N=256")

4 EVALUATION

4.1 Onboard Experiment

To evaluate the performance of the advanced system, the evaluation program using the onboard data used the former paper (Yoo, et al., 2009) which used "T/S Shioji-maru" belong to Tokyo University of Marine Science and Technology was executed. Her principals are shown in Table 1, and the installation of VI-GPS antenna used in this experiment is shown in Figure 6.

Table 2. Principals of T/S Shioji-maru.

Gross Tonnage	425 tons
Lpp	46.00 m
Breadth	10.00 m
Depth	6.10 m
Draught	3.00 m
Displacement	785 DWT
Mean wind direction	165 deg
Mean wind speed	5.3 knots
Data length	1200 s
Sampling freq.	1 Hz

Figure 6. Arrangement on the Flying Bridge of T/S Shioji-maru. (VI-GPS ANT is shown in the white circle)

4.2 Results of Onboard Experiments

To survey the availability of the application of the system using the maneuvering data or onboard wave sensing, onboard experimentation was executed.

STW Information includes some fluctuation, so it was confirmed that conventional Speed Log is not applicable to do with high precise and higher response. It is easily supposed that according to a poor performance of Speed Log it was affected by the wake and/or propeller effects in case of worth installation, etc.

The advanced system modified to add the FFT according to sensing the wave effect. In this experiment original data sampling time is 1 second, so the performance of wave information should be affected by sampling time. So performance of sensing wave and/or current at experimented and ideal are shown in Table 3.

Table 3. Performance of FFT

Items	Experiment Performance	Ideal Performance
Sampling Frequency	1 Hz	5 Hz
Sampling Duration	256 s	204.8 s
Sampling points	256	1024
Frequency Range	1/256 ~ 0.5 Hz	1/102.4 ~ 2.5 Hz
Period Range	2 ~ 256 s	0.4 ~ 102.4 s

The results of wave effect information are shown in Figure 7. Sampling Points of FFT used as shown in the right column of Table 3 was 256, so the resolution of frequency is higher in short period. Figure 7 were recalculated and reformed the period width is approximately 1 sec. In shorter period or higher frequency range will be averaged and gain the accuracy, so the results are shown under the maximum period 17 sec.

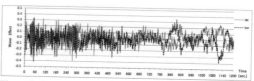

Figure 7. Current and Wave Information on board Experiment.

The sea area of the executed experiments was off Tateyama at the East side of Tokyo Bay entrance, the course of T/S Shioji-maru was approximately 040 degrees including changing course to 030 degrees at 900 seconds. The observed wave information during the experiments was "light breeze", but no recorded of the wave direction.

The solutions of wave effect are shown in Figure 8. The left graphs show wave level of which the unit is meter same as wave height, and lower ones show wave direction which is relative motion to ship not true motion. The parameter of the horizontal axis of all of graphs is the period of encounter.

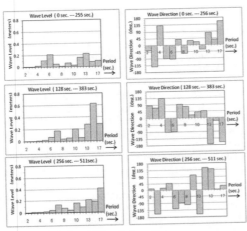

Figure 8. Solution of Wave Information.

Figure 9 shows the time history of wave spectrum for approximately 1,000 sec., so it presents the trend of wave information.

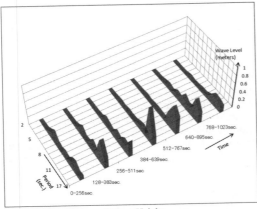

Figure 9. Time History of Wave Height.

Summarize of the results of wave information is following:

1 In early time the wave of which period is 6 sec. and relative direction is -90 degrees. (true direction is 320 degrees).
2 After (1) the true direction of the wave (T=7 seconds) is 080 ~ 130 degrees.
3 The wave (T=6 seconds) and wave (T=7 seconds) fade out after approx. 600 seconds.
4 The wave (T=15 seconds) appears at 2nd duration (128 ~ 383 seconds) and 4th, the maximum level is 0.8 meters and true direction is 030 ~ 045 degrees.

The performance of onboard sensing wave information such as level and direction in these experiments is not able to be close investigated, but the possibility of onboard sensing wave information even during sailing because the trend of wave information sensed by this algorism is reasonable which is wind condition (wind speed: 5 knots, direction: 165degrees), sea state is light breeze, and according to her sailing area the possibility of shadow effect of Peninsula Bousou.

5 CONCLUSION

We discussed the availability of Onboard Wave Sensing using VI-GPS and advanced algorism is proposed to sense the wave effect.

According to the onboard experiments which show good results and possibility to wave effect on board, the possibility to achieve the measurement and application with a good solution of onboard sensing direction of wave with the advanced algorism and the performance of SDME required not only high accuracy but also high response and reliability are shown in this paper.

Onboard Wave Sensing system is also of advantage to monitor the ocean wave using free-mooring buoy with VI-GPS.

Proceeding to develop this system, few points which should be resolved are following:
1 It is very difficult that absolute onboard evaluation would be done, but many cases should be surveyed, and gain the reliability of this system.
2 It is essential to review the response function of ship's motion because of reliability of wave information.
3 To apply the good performance of VI-GPS, one of which is a good response, so using a compass, or Gyro-compass, etc., it takes care to match timing exactly.

REFERENCES

Arai Y., Pedersen E., Kouguchi N. & Yamada K. 2010. The Availability of VI-GPS for Ship-Operation, *In Proceedings of ANC 2010 in Inchon, KOREA*, pp.88-95, November 2010.

Arai Y., Okuda S., Hori A. & Yamada K. 2009. Strategic Application of Two Axes Velocities Information for Ship Maneuvering, *In Proceedings of IAIN Congress 2009 in Stockholm.*

Arai, Y., Okamoto Y., Asaki K., Kouguchi N., & Yonezawa Y. 1990. A Method of Evaluating Simulation on the Ship's Maneuvering, In *Review of the Marine Technical College, No.33*, pp.17-34., March 1990.

Arai, Y., Yonezawa, Y., Kouguchi, N., Hashimoto, S., Yamada, T., Ueda, H. & Nagao, S. 1983. The Measurement of Ship's Path Using the Computer Plotting System with Doppler Sonar. In *Journal of Japan Institute of Navigation*, Vol.68, pp.25-36.

Okuda, S., Arai, Y., Hori A. & Yamada, K. 2008. Simulation Study on the Reliability of Ship's Velocity for Docking Maneuvering. In *Proceedings of 8th Asian Conference on Marine Simulator and Simulation Research*, pp.217—221.

Tatsumi, K., Kouguchi, N., Yoo, Y., Kubota, T. & Arai, Y. 2009. Precise 3-D Vessel Velocity Measurement for Docking and Anchoring. In *Proceedings of International Offshore and Polar Engineering Conference (ISOPE) in Vancouver.*

Tatsumi, K., Fujii, H., Kubota, T., Okuda, S., Arai, Y., Kouguchi, N., & Yamada. K. 2006. Performance Requirement of Ship's Speed in Docking/Anchoring Maneuvering. In *Proceedings of International Association Institute of Navigation IAIN 06, Jeju South Korea*, pp.67-73.

Yoo, Y., Pedersen, E., Tatsumi, K., Kouguchi, N. and Arai, Y. 2009. Application of 3-D Velocity Measurement of Vessel by VI-GPS for STS Lightering. In *International Navigational Symposium on Marine Navigation and Safety of Sea Transportation (TRANS-NAV) in Gdynia.*

4. EGNOS Performance Improvement in Southern Latitudes

L. Panagiotopoulou & K. Frangos
Geotopos S.A., Athens, Greece

ABSTRACT: This paper intends to provide results from "EGNOS Performance Improvement in Southern Latitudes" (EPISOL) project. EPISOL is performed by the Greek company GEOTOPOS S.A. under a contract with the European Space Agency (ESA). EPISOL project aims at analysing, testing and validating the European SBAS EGNOS (European Geostationary Navigation Overlay Service) performance, by outpointing advantages and limitations. It has been designed and operated in a very complicated and demanding environment, the Aegean Sea (Greece, Southern Europe), in which a huge amount of commercial and cruise vessel routes are scheduled daily. Technically, EPISOL also exploits the possibility of EGNOS data collection through other means than the direct Signal in Space (SiS), such as SiSNet (Signal in Space through interNet). Results from this project will form a solid basis towards navigation service improvements and safety enhancements for highly demanded maritime applications, providing important information about EGNOS performance at the edge of the system's service area. In this frame, EPISOL includes a significant number of trials and collection of a large amount of data on coasting vessels in the Aegean before and after the operation of EGNOS Ranging and Integrity Monitoring Station (RIMS) in Athens. As EGNOS data analysis illustrates the European SBAS performance, arguably well-established GNSS navigation techniques, such as GPS RTK, offer reference trajectories for direct comparisons on the position domain.

1 INTRODUCTION

EPISOL has been operated in two Phases: Phase 1 in 2008 before Athens RIMS installation and Phase 2 in 2010 after its deployment and integration in EGNOS ground station network. It has been designed and operated in the Aegean Sea to provide important information about EGNOS performance at the edge of the system's service area.

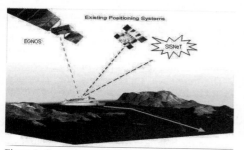

Figure 1: Concept of EPISOL project.

More specifically, EPISOL main objective is the validation of EGNOS relative position accuracy achieved in the Aegean Sea and, in the sequel, the demonstration of alternative methods for redistribution of EGNOS messages in order to overcome EGNOS SiS coverage limitations. In this frame, results from both phases concerning system performance and conclusions for system improvement and future applications in the area, have been drawn.

2 EXPERIMENTS DESIGN

As noted, EGNOS performance is mainly related to the achieved accuracy on the position domain. Therefore, the project has included a significant number of trials and collection of a large amount of data on vessels that sail towards very popular island destinations of the national cabotage, tactically. All routes were carefully chosen in reference with the highly demanding environment of the Aegean Sea and designed for trials in the open sea, as well as for trials for canal, coast and port approach navigation. Additionally, EPISOL analysis presents the achievable system integrity performance in Greece adapting EGNOS standards to International Maritime Organisation (IMO) requirements.

EPISOL analysis also illustrates the continuity of EGNOS SiS and the need to complement with other means of signal transmission. SiSNet combines the powerful capabilities of SBAS navigation and web technologies and thus, EGNOS SiS messages are transmitted via the internet in real time. Figure 1 shows the concept of EPISOL project.

To validate EGNOS performance, the recently established Hellenic Positioning System (HEPOS) has provided GPS RTK reference trajectories with respect to HEPOS network coverage and HEPOS NTRIP RTCM corrections transmission due to the local GPRS network coverage limitations.

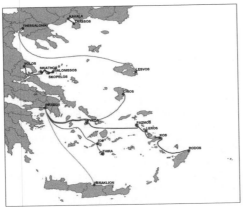

Map 1: EPISOL trials in the Aegean Sea.

Considering these limitations and in accordance with the project's demands, Map 1 shows seven routes to famous Greek islands, that cover a major part of the Aegean Sea, selected to carry out EPISOL trials.

3 DATA COLLECTION PLATFORM ARCHITECTURE

EPISOL data collection platform is described in Figure 2:

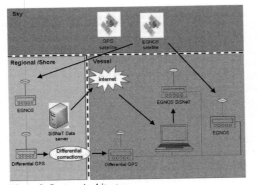

Figure 2: System Architecture.

Two individual Septentrio Polarx2e_SBAS GNSS dual frequency (L1/L2) receivers were installed on board. The first receiver was logging SBAS messages transmitted from both PRN 120 and PRN 126 EGNOS geostationary satellites. PRN 120 broadcasts EGNOS Operational Signal which provides the fully tested system service and PRN 126 broadcasts EGNOS Test Signal, including the latest healthy Athens RIMS data in the system's status configuration. Currently, Athens RIMS is gradually integrated in the system network and the latest system status is continuously tested before its official broadcast. The second receiver accessed EGNOS messages exclusively through SiSNet and the SBAS PVT (Position – Velocity – Time) solution was being internally calculated by the receiver's software. In order to avoid lever-arm effects, both receivers were receiving satellite data from one antenna and an antenna splitter was splitting the signal to the receiver antenna ports. Finally, two laptops connected to a 3G/GPRS modem were offering internet access, providing HEPOS RTCM corrections for the reference trajectories and EGNOS messages through SiSNet server when GPRS network was available. The data collection period of Phase 1 opened at early May 2008 and it was closed at mid July 2008, of Phase 2 opened at May 2010 and it was closed at mid October 2010 while in each Phase almost 70 hours of GNSS/SBAS measurements at 1 Hz rate have been recorded.

4 PERFORMANCE ANALYSIS AND EVALUATION

For the scope of this paper, positioning results using EGNOS from three different routes are displayed, considering the criterion of the equal geographical distribution along the Aegean Sea. Thus, the northernmost Route I, the Route G at the central latitudes of the Hellenic sea area and the southernmost Route A are selected. The performance analysis is focused on different evaluation objectives on the position domain. The main objective that is common to all selected routes, is the comparison of the performance of the achieved position accuracy for both EGNOS Operational Signal and EGNOS Test Signal as transmitted from PRN 120 and PRN 126 respectively. The reference trajectory is the provided HEPOS RTK PVT solution, as long as the vessel was sailing within the limits of HEPOS and local GPRS network coverage. All positioning results from the selected data sets are compared with IMO requirements for both accuracy alone and accuracy / integrity, as well.

Table 1: IMO requirements.

Navigation type application	System / Service level parameters			
	Absolute accuracy	Integrity		
	Horizontal (m)	Protection level (m)	Alarm time (sec)	Availability per 30 days (%)
Ocean / Coastal	10	25	10	99.8
Port approach	10	25	10	99.8
Port	1	2.5	10	99.8

Table 1 shows IMO requirements for different types of navigation applications. The system's integrity level is defined from the calculated position protection limits. Positioning results for EGNOS Operational Signal from Phase 1 and from Phase 2 are illustrated in the analysis performance of Route G. Finally, positioning from EGNOS SiSNet is provided and comparisons between EGNOS SiS Test Signal results and the relative SiSNet results are displayed in the analysis performance of Route I.

5 EGNOS POSITIONING IN SOUTH LATITUDES: RESULTS

Route A

Table 2: Route A EGNOS Performance on the position domain.

Route A Heraklio Piraeus	Performance on the position domain			
	VPE EGNOS Test	HPE EGNOS Test	VPE EGNOS	HPE EGNOS
Mean (m)	0.20	2.11	1.05	1.64
Standard Deviation (m)	2.26	3.47	5.29	2.08
2-sigma 95% (m)	4.71	9.05	11.62	5.80
Availability (IMO Req) %	73.0		12.5	

As noted, EGNOS Horizontal Position Errors (HPE) and EGNOS Vertical Position Errors (VPE) that visualize system's accuracy performance are calculated using reference position the RTK PVT solution provided by HEPOS. It is remarked that HEPOS reference position accuracy is perturbed by all factors that influence RTK positioning. However, RTK method (when available) offers considerably, the optimum navigation solution and especially in maritime applications it is ideally used for position error calculations. Table 2 shows system's accuracy performance for both EGNOS Test and Operational Status. The general comment from this Table is that EGNOS Test signal mean values are more than 5 times less than the corresponding values of EGNOS Operational and the standard deviation on the vertical direction (height accuracy performance) is 135%

improved. Nevertheless, EGNOS Operational Signal mean values are 25% improved in comparison with the corresponding values of the Test signal and the standard deviation is almost 40% improved horizontally (with relevance to the corresponding RTK solutions).

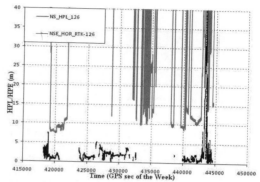

Figure 3: Route A EGNOS Test HPE/HPL time series.

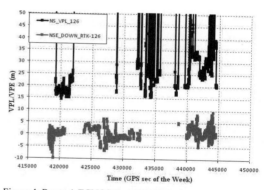

Figure 4: Route A EGNOS Test VPE/VPL time series.

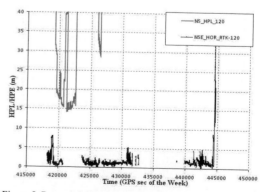

Figure 5: Route A EGNOS Operational HPE/HPL time series.

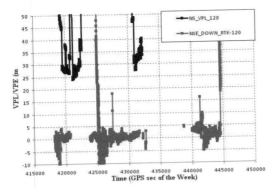

Figure 6: Route A EGNOS Test VPE/VPL time series.

Figures 3-6 show the time series of EGNOS HPE and VPE along with the Horizontal Protection Limits (HPL) and the Vertical Protection Limits (VPL). Figures 3 and 4 correspond to the EGNOS Test signal time series plots. On Figure 3 the grey dotted line represent the HP limit and the black dot represents the horizontal position error whereas, on Figure 4 the black dotted line represents the VP limit and the grey dots represent the vertical position error. Respectively, Figures 5 and 6 correspond to the EGNOS Operational signal time series plots. It is evident that EGNOS Test signal offers larger time spans of protection limits than EGNOS Operational. Therefore and in accordance with Table 1 concerning IMO requirements for different navigation modes, EGNOS Test delivers significantly better results. Maps 2 and 3 show dynamic plots for EGNOS Operational and Test respectively, corresponding to IMO requirements for accuracy alone. Simple dots are the epochs where HPE is more than 10m, small circles represent epochs where HPE is less than 10m (requirements for ocean, coastal and port approach navigation), while star shapes are epochs that correspond to HPE less than 1m (port navigation).

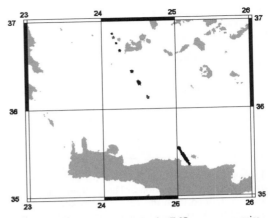

Map 3: Route A EGNOS Test plot for IMO accuracy requirements.

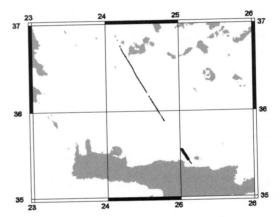

Map 4: Route A EGNOS Operational plot for both accuracy and integrity IMO requirements.

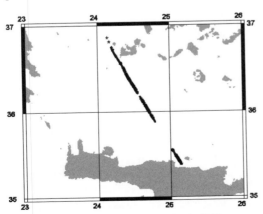

Map 2: Route A EGNOS Operational plot for IMO accuracy requirements.

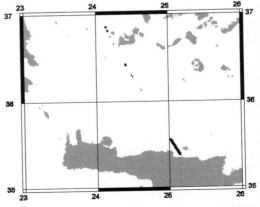

Map 5: Route A EGNOS Test plot for both accuracy and integrity IMO requirements.

Accordingly, Maps 4 and 5 are dynamic plots for EGNOS Operational and EGNOS Test respectively, that correspond to IMO requirements for both accuracy and integrity on different navigation modes. Simple dots represent positions where Horizontal Position Limit value is larger than 25meters. Square shapes are epochs at which the position horizontal limit is less than 25 meters and horizontal position error is less than 10 meters at the same time, conditions that cover IMO requirements for open sea, coastal and port approach navigation. Star shapes represent epochs where HPL is less than 2.5m and Horizontal Position Error is less than 1 meter at the same time. According to Figures 3 and 5 this route has not had positioning results where HPE is less than 1 meters and HPL is less than 2.5 meters at the same time, conditions required for port navigation. However, as shown on Maps 2 and 3, both Test and Operational Signal HPE results meet the IMO requirements for port navigation.

Route G

Table 3: Route G / Phase 1 and 2 EGNOS Performance on the position domain.

Skiathos Volos	Performance on the position domain							
	VPE EGNOS TEST		HPE EGNOS TEST		VPE EGNOS		HPE EGNOS	
Phase	1	2	1	2	1	2	1	2
Average (m)	1.26	0.27	0.89	1.20	1.22	0.30	2.82	2.23
Standard Deviation (m)	2.30	1.56	0.63	0.63	8.88	3.76	15.95	5.21
2-sigma 95% (m)	5.86	3.39	2.47	2.47	18.99	7.22	34.72	12.65
Availability (IMO Req) %	89	98	89	98	16	41	16	41

EGNOS positioning results from Route G have been selected to outline the system's performance improvements on the position domain between the two Phases of the project. Table 3 shows the basic statistics of the results. VPE EGNOS Test mean values is more than 4.5 times improved in Phase 2, whereas the HPE EGNOS Test mean values are 25% improved as well. The VPE EGNOS Test standard deviation is 33% improved in Phase 2 and HPE EGNOS Test standard deviation is on the same level for both phases. Accordingly, VPE EGNOS Operational mean values are extremely improved in Phase 2, whereas the corresponding standard deviation is almost 60% improved as well. Finally, HPE EGNOS Operational mean values are almost on the same level in both phases, when the standard deviation is almost 3 times improved in Phase 2.

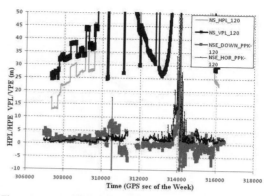

Figure 7: Route G Phase 1 EGNOS Operational HPE/HPL and VPE/VPL time series.

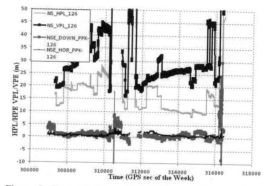

Figure 8: Route G Phase 1 EGNOS Test HPE/HPL and VPE/VPL time series.

Figures 7 and 8 display the horizontal and vertical position errors along with the horizontal and vertical protection limits time series charts in Phase 1. Figure 7 corresponds to the HPE-HPL / VPE-VPL performance through time as provided from EGNOS Operational signal and Figure 8 corresponds to the HPE-HPL / VPE-VPL time series of EGNOS Test signal.

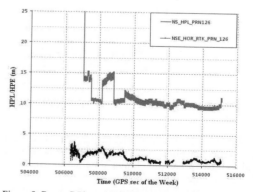

Figure 9: Route G Phase 2 EGNOS Test HPE/HPL time series.

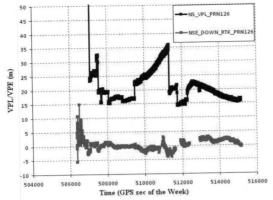

Figure 10: Route G Phase 2 EGNOS Test VPE/VPL time series.

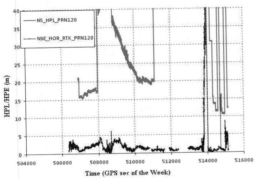

Figure 11: Route G Phase 2 EGNOS Operational HPE/HPL time series.

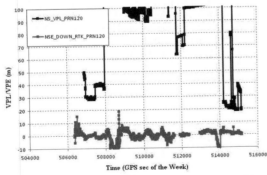

Figure 12: Route G Phase 2 EGNOS Operational VPE/VPL time series.

Accordingly, Figures 9-12 display the horizontal and vertical position errors along with the horizontal and vertical protection limits time series charts in Phase 2 Figures 9 and 10 show HPE-HPL and VPE-VPL diagrams of EGNOS Test, respectively. In the same manner, Figures 11 and 12 show the HPE-HPL and VPE-VPL diagrams of EGNOS Operational Signal. Comparing the integrity performance between the Operational and Test signal, it is obvious that system's integrity performance is enhanced after the installation and deployment of Athens RIMS.

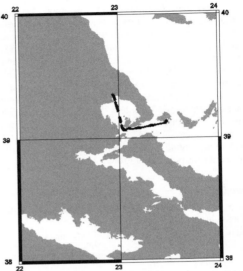

Map 6: Route G Phase 1 EGNOS Operational plot for IMO accuracy requirements.

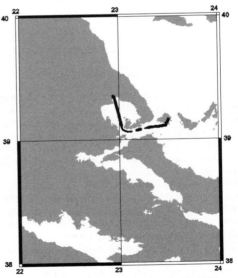

Map 7: Route G Phase 1 EGNOS Test plot for IMO accuracy requirements.

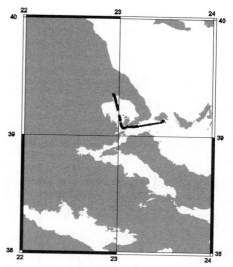

Map 8: Route G Phase 2 EGNOS Operational plot for IMO accuracy requirements.

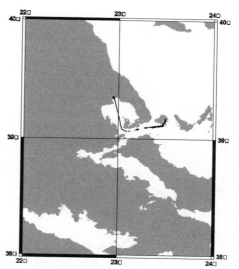

Map 10: Route G Phase 1 EGNOS Operational plot for both accuracy and integrity IMO requirements.

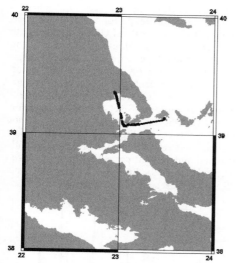

Map 9: Route G Phase 2 EGNOS Test plot for IMO accuracy requirements.

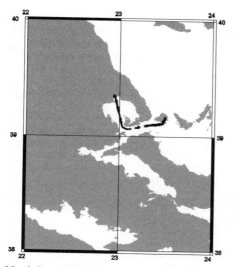

Map 11: Route G Phase 1 EGNOS Test plot for both accuracy and integrity IMO requirements.

Maps 6-9 are the IMO requirements plots for accuracy alone and correspond to EGNOS Test and EGNOS Operational signal performance for both Phase 1 and Phase 2. Same, Maps 10-13 are the IMO requirements plots for both accuracy and integrity for EGNOS Test and EGNOS Operational signal performance, accordingly. A close examination of Map 13 shows that the system performance improvement in both accuracy and integrity during Phase 2 is even clearer.

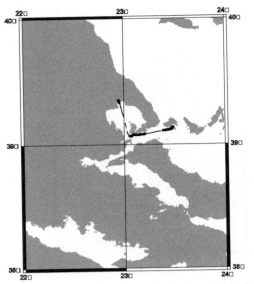

Map 12: Route G Phase 2 EGNOS Operational plot for both accuracy and integrity IMO requirements.

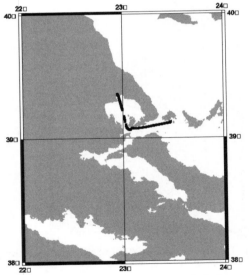

Map 13: Route G Phase 2 EGNOS Test plot for both accuracy and integrity IMO requirements.

Route I

Table 4: Route I - EGNOS SiS and SiSNet Performance on the position domain.

Thasos Kavala	Performance on the position domain							
	VPE EGNOS TEST		HPE EGNOS TEST		VPE EGNOS		HPE EGNOS	
Phase	SiS	SiS NeT	SiS	SiS NeT	SiS	SiS NeT	SiS	SiS NeT
Average (m)	0.13	0.47	1.12	0.47	8.60	1.01	3.63	1.53
Standard Deviation (m)	3.99	4.00	1.54	1.76	11.90	2.96	4.27	0.81
2-sigma 95% (m)	8.12	8.48	4.20	3.98	32.40	6.93	12.16	3.14
Availability (IMO Req) %	97	97	97	97	15	N/C*	15	N/C*

(*N/C – Not Computed)

EGNOS performance analysis in the European south latitudes includes performance comparisons between SiS and SiSNet, aiming at the evaluation of alternative means of receiving EGNOS messages than the direct satellite signal reception. Table 4 is displaying SiS and SiSNet performance on the position domain. It is obvious that EGNOS Test for both message reception methodologies perform alike on the vertical position direction, whereas small differentiations are observed on the horizontal position accuracy. It is also evident, that the levels of the position accuracy for EGNOS Operational are extremely improved using SiSNet, however protection limits for EGNOS SiSNet Operational were not available. For this reason Figures 13 and 14 display the horizontal and vertical position errors along with the horizontal and vertical protection limits time series charts, for EGNOS SiS Test signal alone and EGNOS SiSNet Test signal alone, respectively.

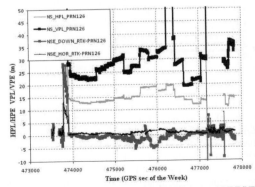

Figure 13: Route I EGNOS SiS Test HPE/HPL and VPE/VPL time series.

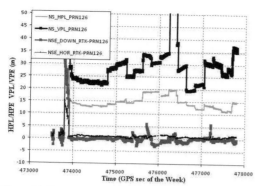

Figure 14: Route I EGNOS SiSNeT Test HPE/HPL VPE/VPL time series.

6 CONCLUSIONS

EPISOL is a project mainly concentrated on the performance analysis of the position domain. Namely, it is focused on the position accuracy and integrity that can be achieved using EGNOS for maritime applications. As the project has taken place in two Phases, before and after Athens RIMS deployment, using SiSNet as the alternative means for EGNOS messages reception, the most important conclusion drawn by the analysis results is the significant system improvement after the RIMS deployment. Actually, and since RIMS data are gradually integrated into the system's new configuration, it is anticipated that EGNOS accuracy and integrity performance at the south latitudes, shall further be improved at the time of the complete integration of the RIMS in the system's network. Moreover, it has been proved that EGNOS SiSNet could equally replace SiS reception in environments and under conditions that SiS reception is not available. Finally, it has been shown that even under the current configuration status, EGNOS can be used as the primer navigation system for many maritime applications meeting IMO requirements for sea navigation.

ACKNOWLEDGEMENTS

The work presented in this paper has been performed and funded by ESA in the framework of a contract under "ESA - Greece Incentive Scheme - 1st Call for Ideas".

REFERENCES

EGNOS Performance Improvement in Southern Latitudes (EP-ISOL), Phase 1 and 2 (Contract 20973), Final report of Phase 1, GEOTOPOS S.A., 2008.

EGNOS Performance Improvement in Southern Latitudes (EP-ISOL), Phase 1 and 2 (Contract 20973), Proposal for Phase 2, GEOTOPOS S.A., 2009.

EGNOS Performance Improvement in Southern Latitudes (EP-ISOL), Phase 1 and 2 (Contract 20973), Draft report on "Data Performance Evaluation" of Phase 2. -GEOTOPOS S.A., 2011.

Revised Maritime Policy and Requirements for a Future Global Navigation Satellite System (GNSS), IMO Resolution A.915(22) adopted on 29 November 2001.

World-wide Radionavigation System, IMO Resolution A.953 (23) adopted on 5 December 2003.

World-wide Radionavigation System. Evaluation of Galileo Performance against Maritime GNSS Requirements, IMO Sub-Committee on Safety of Navigation NAV 49/13, 16 April 2003.

EGNOS Service Definition Document Open Service, Ref: EGN-SDD OS V1.1, European Commission Directorate-General for Energy and Transport.

EGNOS Safety of Life Service Definition Document, Ref: EGN-SDD SoL, V1.0, European Commission Directorate-General for Energy and Transport.

5. An Integrated Vessel Tracking System by Using AIS, Inmarsat and China Beidou Navigation Satellite System

C. Yang, Q. Hu, X. Tu & J. Geng
Merchant Marine College, Shanghai Maritime University

ABSTRACT: As there are more and more Automatic Identification System (AIS) sets have been deployed onboard, it is getting easier for people to trace vessels. Today, many online vessel monitoring services have been developed; however, most of them are based on AIS information. Because the coverage limitation of VHF frequency, which the AIS set works on, is normally no more than 25 nautical miles, so these systems can not track vessels when they are beyond the coverage of the shore-based AIS station. In order to track vessels in all sea areas, we developed a comprehensive vessel tracking system, namely ManyShips, which integrates AIS, Inmarsat and China Beidou navigation satellite system. The running result of the system shows that the Beidou satellite system can track vessels within Asian-Pacific region while the Inmarsat-C station polling service can help people tracking vessels within sea area A3.

1 BACKGROUND

Today, there are more and more base stations have been developed to receive the real-time vessel information and to send them to VTS (Vessel Traffic Service) where this information will be displayed on ECDIS to facilitate traffic monitoring. And the information is very valuable in other area in shipping industry such as ship agent, brokerage, pilotage, salvation, custom inspection, quarantine, and fleet monitoring, etc.[1][2]

However, it is hard for these people access the AIS information field.[3] To satisfy their demand, several live AIS web system have been developed, for example, Lloyds AISLive, NavCom AIS Live, AISLivepool and Tokyo bay live traffic website. All these systems, however, are only based on AIS information which is collected from shore based AIS stations, and their coverage only within sea area A1. We have developed a live AIS web system using Ajax technology[4][5]. But this system can not track vessels when they are beyond the coverage of shore-based AIS stations, because the coverage of the VHF frequency, which the AIS system works on, is normally within sea area A1. Therefore, in this paper we integrate Inmarsat-C polling service and China Beidou Navigation satellite system into the vessel tracking system in order to extend coverage of the system to sea area A3. Once the vessel leaves the sea area A1, that means the position information broadcasted by AIS system can not be received by the shore-based AIS station, the vessel tracking system can use Inmarsat-C polling services or China Beidou navigation satellite system to tracking the vessel continuously.

This paper is organized as follows: in section 2, we briefly introduce the AIS system, Inmarsat-C polling service and China Beidou navigation satellite system; in section 3, we describe the architecture of the vessel tracking system; in section 4 we give out the running results and conclusions.

2 INTRODUCTION

2.1 *The Automatic Identification System*

The Automatic Identification System is an automated tracking system used on vessels and Vessel Traffic Services (VTS) for identifying and locating vessels by electronically exchanging data with other nearby ships or VTS stations. AIS integrates a standard VHF transceiver with a positioning system such as LORAN-C or GPS receiver that can provides information supplements the marine radar, which is the primary method of collision avoidance for water transportation. The AIS provides many important information such as unique identification number, namely Maritime Mobile Service Identification number (MMSI in short), position, course, speed and rate of turn, and can be displayed on a screen or an ECDIS.[6][7][8]

2.2 The Inmarsat-C polling service

The coverage of Inmarsat is shown in fig. 1.

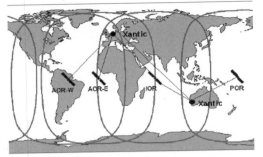

Figure 1[9]. The coverage of Inmarsat.

Data reporting and polling are value added services based on shipborne Inmarsat-C terminal. The data reporting services is intended for transferring small quantities of data (e.g. a position report) from an Inmarsat-C terminal to a predetermined address. This predetermined address could be an internet email address, a telex, a telephone-modem and so on. Data reports make efficient use of the Inmarsat-C system. Data packets limited to a maximum of 256 bits （32 bytes） are transmitted via signaling channels of the Inmarsat-C network. Time and cost are saved by avoiding switching to a messaging channel. Data report can be sent directly from a C-terminal or command with a poll. Most C-terminal can transmit Data Reports manually by means of an operator or be programmed for automatic transmission at pre-set intervals. And the same can be achieved from a remote location (e.g. a fleet management system) by sending a Poll to the C-terminal commanding the sending of Data Report. A Poll is a short command to an individual C-terminal or group of C-terminals initiating some action, controlled by the software of the C-terminal. A fleet manager can ask for data reports, with for instance the position of his ships. Polls can be sent via Internet e-mail. The polling service process procedure is show in fig. 2 and 3.[9]

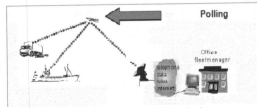

Figure 2. Polling the mobile Inmarsat-C terminals.

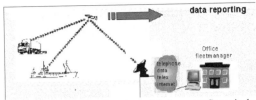

Figure 3. Data reporting from the mobile Inmarsat-C terminals.

2.3 The China Beidou navigation satellite system

China Beidou navigation satellite system, which is developed by China stand-alone, is active three-dimensional satellite positioning and communication system. The system can provide positioning and navigation service, time service and communication service. Now it is in its 1st generation stage so the coverage is from 5° N to 55°N, 70° E to 140° E. (See fig. 4) The system consists of satellites, ground earth stations and user side. There are 5 geostationary earth orbit (GEO) satellites and 30 non-GEO satellites. The ground earth stations including control stations, upload stations and monitoring stations. The user side is a receiver which is compatible with GPS, GLONASS and GALILEO. According to the implementation plan of the China Beidou navigation satellite system, it can provide global service at 2020.[10]

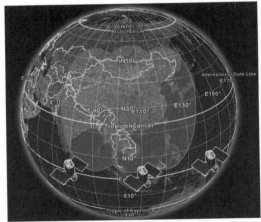

Figure 4. the coverage of China Beidou navigation satellite system.

3 SYSTEM ARCHITECTURE

The vessel tracking system we developed, namely ManyShips, integrates the AIS, Inmarsat-C and China Beidou satellite system. The system architecture is shown in fig.5.

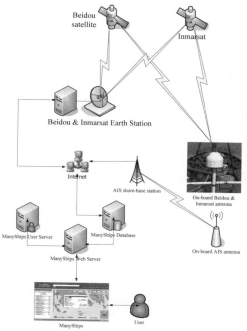

Figure 5. ManyShips system architecture.

When the vessel is navigating within sea area A1, the shore-based AIS station can collect the information broadcasted from the on-board AIS system. Once the vessel is leaving sea area A1, the on-board Beidou satellite antenna can send out the vessel's position information up to the Beidou satellite actively, then the information could be forwarded to the ground earth station and ManyShips system receives it via the Internet finally. Because the Beidou navigation satellite system now only has a regional coverage, so users should send out polling command from ManyShips system to the on-board Inmarsat-C terminals to request data report via the Inmarsat system when the vessel is out of the Beidou coverage.

There is little different between the Beidou satellite report procedure and that of the Inmarsat-C polling report. The on-board Beidou antenna can send out data report up to the Beidou satellite actively and the data report will be pushed to ManyShips system by the ground earth station will the station receiving it from the satellite. So the ManyShips system can locate the vessels continuously when the vessel navigating within the coverage of the Beidou navigation satellite system. While the data report from the onboard Inmarsat-C terminals is initiated when the terminal receives the polling command sent from shore users. The shore users send out an email, whose subject is just the polling command, to the ground earth station, and then the polling command will be uploaded to the Inmarsat system by the station. Also the data report is returned to the users in

email format sent by the ground earth station once the data report from the vessels is received.

4 RUNNING RESULT AND CONCLUSIONS

We have deployed many shore-based AIS stations among China's sea line. And these stations can collect the vessels' AIS information when they are navigating within the coastal area. Fig. 6 shows the shore-based AIS stations coverage with the green points stand for the vessels.[11]

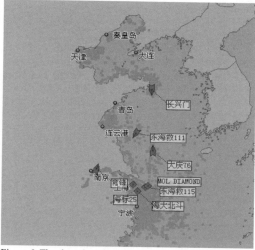

Figure 6. The shore-based AIS stations coverage

A Beidou navigation satellite antenna has been deployed onboard M/V Yu Feng, which sails between Nanjing and Busan. Fig. 7 shows its track when she departures Gwangyang.

Figure 7. Track information reported from Beidou satellite system.

All the position information is reported from the on-board Beidou antenna to the ManyShips system every 15 minutes actively.

We have also resisted the Inmarsat-C polling service on M/V QiLinZuo, a vessel that belongs to China Shipping Company and navigating between Shanghai China and European ports where we have not deployed shore-based AIS stations and also out of the coverage of China Beidou navigation satellite system, so it likes the vessel is navigating in sea area A3. We have sent out several polling commands to the vessel when she is navigating in Aegean Sea. Fig.8 shows the vessel's track which is generated from its Inmarsat-C data reports.

Figure 8. Track information reported from Inmarsat-C station polling service.

The above running results show that the China Beidou navigation satellite system and Inmarsat-C polling service are very important components when developing the integraed vessel tracking system. They can extend the coverage of the ManyShips system from sea area A1 to sea area A3

ACKNOWLEDGMENT

This research is supported by Shanghai Education Committee with grant No.08YZ107, Science and Technology Program of Shanghai Maritime University with grant No.20100134 and the 2010 Shanghai Education Committee dedicated fund for selection and training of scientific research for outstanding young teachers.

REFERENCES

[1] Wang Mingshi and Zhang Renying,(2006). "Analysis of business value of AIS in shipping logistics industry," Port Science and Technology (in Chinese), pp.50-51.
[2] Liao Yifan, (2006). "Application of AIS in ship brokerage business," Warter Transportation Management, Vol.28, No.5.
[3] Hu Qinyou, Yang Chun, and Shi Chaojian, "Portlive, A Bridge to Ship AIS Information Island," Proceedings of the Eighth International Conference of Chinese Logistics and Transportation Professionals vol. 3, pp. 1922–1927.
[4] Hu Qinyou, Chen Jinhai, and Shi Chaojian, "Bring Live AIS Information on the Web Charts by Using Ajax," Proceedings of the 7th International Conference on Intelligent Transport Systems Telecommunications, pp.455–459.
[5] Chun Yang, Qinyou Hu et al., "Active Vessel Navigation Monitoring with Multi-media Message Service," Proceedings of the 2nd International Conference on Future Generation Communication and Networking, pp. 1–13.
[6] International Association of Maritime Aids to Navigation and Lighthouse Authorities (IALA), (2002). "IALA guidelines on the universal automatic identification system (AIS), vol. I, Part II, Technical Aspects of AIS," Edition 1.1.
[7] International Telecommunication Union, ITU-R Recommendation M. 1371-1, (2001). "Technical characteristics for an universal shipborne automatic identification system using time division multiple access in the VHF marine mobile band."
[8] International Electrotechnical Commisioin. IEC 6993-2, (2001). Maritime navigation and radio communication requirments – automatic identification systems (AIS) – part 2: class a shipborne equipment of the universal automatic identification system (AIS) – operational and performance requirements, methods of test and required test results. Edition 1.
[9] http://www.xantic.net
[10] http://www.beidou.gov.cn
[11] http://www.manyships.com

Positioning Systems

6. Recent Advances in Wide Area Real-Time Precise Positioning

D. Łapucha
Fugro Chance, Lafayette, USA

K. de Jong & X. Liu
Fugro Intersite, Leidschendam, The Netherlands

T. Melgard, O. Oerpen & E. Vigen
Fugro Seastar, Oslo, Norway

ABSTRACT: This paper describes briefly new high precision wide area real time positioning systems, discusses their evolution, implementation and presents recent results. In the latter part the paper is primarily focused on discussing the benefits of GLONASS augmented high precision positioning.

1 INTRODUCTION

Differential GPS (DGPS) has been an established technique that provides meter level positioning in real time over wide area typically using single frequency Global Navigation Satellite System (GNSS) receivers. DGPS systems may utilize single reference station based corrections, such as IALA maritime stations, or orbit and clock corrections derived from wide area network, such as US WAAS and European EGNOS. The accuracy of DGPS technique is limited to about one meter level because of inherent pseudorange noise. Also, single frequency based DGPS accuracy may deteriorate during ionospheric storms.

Real Time Kinematic (RTK) positioning technique can potentially provide centimeter-level positioning accuracy. However, RTK requires the distance to the closest reference station to be within several tens of kilometers. Because of its fundamental distance limitations RTK is not well suited for wide area and offshore positioning.

To address the needs of higher accuracy users Fugro has pioneered over the last ten years several decimeter level wide area real time positioning systems, called HP, XP, and the most recent G2. The XP and HP systems are based on the use of GPS satellites only, while G2 is augmented with GLONASS satellites. G2 is the first real-time positioning system combining observations from GPS and GLONASS that has wide area coverage. The addition of GLONASS satellites in G2 positioning solutions results in improved availability and robustness of high precision positioning compared to HP and XP, GPS only systems.

These real-time systems developed by Fugro achieve decimeter accuracy by using dual frequency carrier phase observations to eliminate ionospheric effects and provide high accuracy positioning. By offering decimeter level accuracy over wide area, Fugro high accuracy systems have bridged the accuracy and coverage gaps between a confined area centimeter level RTK and a wide area meter level DGPS.

Fugro high precision services use wide footprint area geostationary satellites to transmit corrections determined from global reference networks to the GNSS users, conceptually shown in Figure 1. Fugro developed HP, XP and G2 solutions are embedded in virtually thousands of GNSS receivers. These receivers may be enabled to receive the corrections broadcast via geostationary satellites. The receivers with Fugro enabled HP, XP and G2 high precision solutions are used in various applications such as Dynamic Positioning of drill ships/supply vessels, airborne lidar surveys, hydrographic surveys, tidal control and precision agriculture.

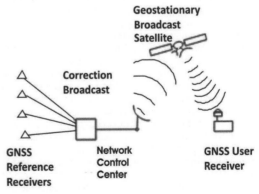

Figure 1. Geostationary Satellite Correction Broadcast

This paper describes briefly Fugro high precision systems, discusses their evolution, implementation and presents recent results. In the latter part the paper is primarily focused on discussing the benefits of combined GPS and GLONASS positioning as offered by G2 system.

2 CARRIER PHASE POSITIONING

Unlike standard DGPS systems that use lower precision pseudorange observations Fugro high accuracy systems use higher precision dual frequency carrier phase data as primary observations in a positioning solution. The use of dual frequency carrier phase observations enables virtual eliminating of ionospheric delays, one of the major error sources of GNSS positioning. The carrier phase observations are adjusted with the respective HP, XP or G2 correction information before entering the user positioning filter. Carrier phase observations are also corrected for various disturbing effects using high fidelity error modeling.

The satellite ambiguities, inherent in carrier phase observations, are recomputed for the rising satellites that are brought in a solution. The carrier phase ambiguities and residual atmospheric effects are then estimated, along with a position, primary parameter of interest in a user solution.

Achieving high accuracy after the cold start requires some initial convergence time, typically 10 to 30 minutes. The convergence time is a function of satellite geometry and tracking environment. Generally, shorter convergence time is achieved with better satellite geometry.

3 MULTI REFERENCE HP

Fugro introduced the first commercial decimeter level wide area real time positioning system in 2001. This service named HP, is based on the application of virtual station corrections optimized for the user location (Lapucha et al, 2001). The virtual corrections are computed using the corrections from the region of the multiple reference stations typically close to or surrounding the user. The virtual base station corrections are applied to the rover observations to mitigate satellite clock and orbit errors.

Fugro operates a worldwide network of about a hundred HP reference stations providing high precision coverage for major land masses and coastal areas. The HP solution approaches RTK accuracy if the user is within the reference network and the closest station is generally less than 1000 km. However, HP positioning accuracy gradually deteriorates if the user is outside of the network and the distance to the closest station is more than 1000 km. Thus the HP system is not suitable to provide high accuracy posi-

tioning in truly remote areas from reference stations such as in the middle of the oceans.

Typical 24 hour monitoring results from the HP system operating in dynamic mode at Lafayette, USA, on the Gulf of Mexico coast, using four reference stations with the distances ranging from 350 km to more than 1000 km, observed on January 23, 2011 are shown in Figure 2. The position accuracy given in terms of standard deviation is 3, 3 and 5 cm, for longitude, latitude and height, respectively.

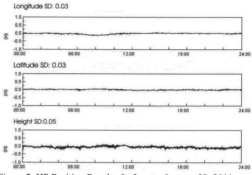

Figure 2. HP Position Results, Lafayette, January 23, 2011

HP solution was also extensively tested at various locations and over an extended time. The results of these tests showed that HP solution provides 10 cm horizontal, 15 cm vertical accuracies 95% at distances up to 500 km and 15 cm horizontal, 30 cm vertical accuracies 95% at distances up to 1000 km from the closest reference station.

4 PRECISE POINT POSITIONING XP

The XP system, based on the Precise Point Positioning (PPP) method, was introduced by Fugro in 2003. The precise orbit and clock corrections used in the XP system are determined in cooperation with National Aeronautics and Space Administration (NASA) Jet Propulsion Laboratory (JPL) based on NASA's worldwide network of reference stations.

The PPP method used in the XP system involves using the satellite specific precise orbit and clock corrections instead of virtual reference station range corrections. These corrections represent the most accurate estimate of the errors of the GNSS satellites broadcast orbit and clocks. Unlike the corrections used in the multiple reference station HP method, which are reference station and satellite specific, the orbit and clock corrections used in the PPP XP method are satellite specific only and not location dependent. The positioning accuracy is no longer limited by the distance from the reference stations. Therefore, application of these corrections leads to virtually homogeneous high positioning accuracy worldwide.

Typical 24 hour monitoring results from the XP system operating in dynamic mode at Lafayette, USA, on the Gulf of Mexico coast, as observed on January 23, 2011 are shown in Figure 3. The position accuracy given in terms of standard deviation is 4, 3, and 8 cm, for longitude, latitude and height, respectively.

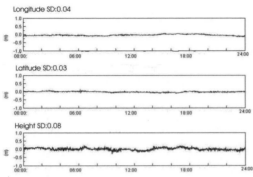

Figure 3. XP Position Results, Lafayette, January 23, 2011

XP solution was also extensively tested at various locations and over an extended time. The results of these tests showed that the XP solution provides 10 cm horizontal, 20 cm vertical accuracies in terms of 95% statistics. Unlike HP, the accuracy of XP is not dependent on location and distance from the reference stations.

5 GLONASS AUGMENTED PRECISE POINT POSITIONING G2

Fugro introduced in 2009 truly the next generation multi constellation real-time PPP system, based on the use of precise GPS and GLONASS orbit and clock corrections, called G2 (Melgard et al, 2009). The development has benefited from the close cooperation between Fugro and the European Space Operation Centre (ESOC), an establishment of the European Space Agency (ESA). ESOC has contributed with their expertise on precise orbit and clock processing techniques while Fugro built an operational real time system.

G2 position solution uses the PPP method with fine tuned statistical models to process GPS and GLONASS satellite observations and precise orbits and clock corrections determined from the global G2 network. These corrections are satellite specific only and not location dependent, similarly to XP. Therefore, the application of these corrections in a G2 user solution leads to virtually homogeneous high positioning accuracy worldwide

The G2 service utilizes Fugro's network of dual system GNSS reference stations to calculate precise orbits and clocks on a satellite by satellite basis for all 50 plus satellites of the two global navigation satellite systems. The system comprises about 40 dual-frequency GPS and GLONASS reference stations, operated independently of HP and JPL networks, evenly distributed around the world.

Successful integration of GLONASS carrier phase observations in G2 solution required accounting for incompatibilities between GPS and GLONASS systems. GLONASS satellites, unlike GPS, use different satellite specific frequencies. Also, GLONASS observations refer to different time system than GPS. However, after accounting for these differences GLONASS satellites act like additional GPS satellites in G2 solution.

Including GLONASS together with GPS satellites improves redundancy, geometry and availability of a positioning solution. Because of the greater number of satellites and improved geometry, integrated GPS and GLONASS G2 solution offers faster convergence than GPS only solution (Melgard et.al, 2009). Additional GLONASS satellites offer the potential to enable a positioning solution that may not be possible with GPS only, especially in challenging tracking environments with line of sight obstructions such as depicted in Figure 4.

Figure 4. Challenging GNSS Tracking Environment

6 G2 POSITION ACCURACY ANALYSIS

In the following, G2 positioning results are presented from different locations and times to assess the representative G2 accuracy figures. These results were achieved with the systems operating in dynamic mode. It should be noted, the daily positioning accuracy can vary from day to day and with location depending on GNSS receiver, antenna and antenna cable and local environment. It is therefore important to note the following results represent examples observed at the monitor stations.

Example, 24 hour monitoring results time series from G2 system operating at Gulf of Mexico, Lafayette location are shown in Figure 5. The position accuracy given in terms of standard deviation is 2, 2 and 6 cm, for longitude, latitude and height, respectively.

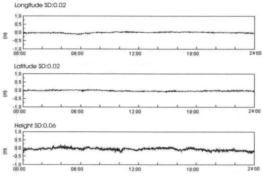

Figure 5. G2 Position Results, Lafayette, January 23, 2011

Example, monitoring results from the G2 system operating in Oslo, Norway are shown in Figure 6. The position accuracy given in terms of standard deviation is 2, 2 and 4 cm, for longitude, latitude and height, respectively.

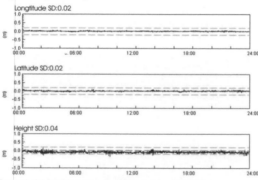

Figure 6. G2 Position Results, Oslo, January 2, 2011

Similar data were also collected in different locations around the world. Composite G2 accuracy figures given in terms of 95% statistics from some locations on January 23, 2011 are summarized in Figure 7. These results demonstrate that the G2 solution provides consistently, even with incomplete GLONASS constellation, 10 cm horizontal and 15 cm vertical accuracies in terms of 95% statistics.

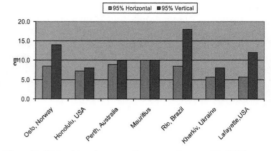

Figure 7. G2 Position Accuracy Summary, January 23, 2011

7 BLOCKAGE SIMULATION

Combining GPS and GLONASS observations in the integrated G2 solution offers potential to expand availability of high precision solution when single satellite system solution is not possible, as shown earlier at Figure 4. Generally, at least four satellites are necessary for single satellite system three dimensional user position determination.

To assess G2 performance under blockage conditions G2 data was reprocessed with simulated virtual wall to the south blocking GNSS satellite observations, as shown in Figure 8.

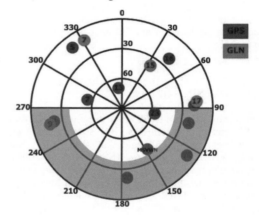

Figure 8. Satellite Blockage Skyplot

During the simulated blockage times, when a number of GPS satellites drops below the required four, GPS only solution fails to provide position, as shown in Figure 9. Moreover, positioning accuracy deteriorates after outage because of reconvergence in poor satellite geometry conditions.

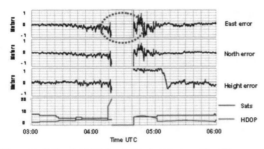

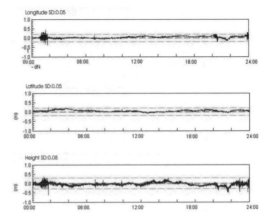

Figure 9. GPS only PPP Results Under Blockage Conditions

However, during time when GPS solution was not available, G2 solution provided seamlessly decimeter level positioning with only slightly degraded accuracy during blockage time, as can be in Figure 10. These results demonstrate improved resiliency of G2 solution in challenging tracking environment.

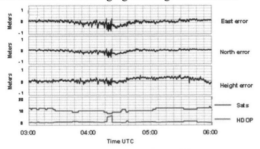

Figure 10. G2 Results Under Blockage Conditions

8 GLONASS ONLY PPP RESULTS

Expanding GLONASS constellation offers potential for PPP solution independent on GPS. As of beginning of 2011, GLONASS constellation does not offer 24 hour worldwide coverage. However it provides virtually 24 hour coverage of Northern Europe.

Figure 11 presents recent position results from the PPP system in Oslo utilizing only GLONASS observations and precise corrections in a user solution. Moreover, GLONASS precise orbit and clock corrections used in a solution were determined in the process that also used GLONASS observations only, completely independent on GPS.

The GLONASS PPP position accuracy from this test given in terms of standard deviation is 5, 5 and 8 cm, for longitude, latitude and height, respectively. Even with incomplete GLONASS constellation, these results demonstrate that GLONASS PPP method can potentially provide decimeter accuracy positioning independent on GPS.

Figure 11. GLONASS PPP Results, Oslo, January 26, 2011

9 OPERATIONAL SERVICES

Fugro HP, XP and G2 services and networks are operated independently and redundantly to provide the highest level of positioning integrity. All Fugro positioning services use geostationary satellites transmitting data within L band, the same frequency band as used by GPS and GLONASS satellites. Use of L band has the advantage that user receivers can utilize the same antenna for reception of GNSS and geostationary satellite signals. The geostationary satellites used by Fugro provide virtually worldwide coverage, with the exception of polar regions, as can be seen in Figure 12.

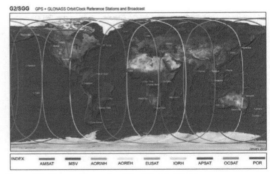

Figure 12. Geostationary Satellite Worldwide Coverage

Fugro developed HP, XP and G2 solutions are embedded in major brand GNSS receivers. The respective solutions can be activated with a subscription. These receivers employ the Fugro software library that decodes subscription and correction data and carries out high precision positioning computations. HP, XP and G2 enabled GNSS receivers are used for numerous applications requiring high precision positioning on land, sea and in the air.

10 CONCLUSIONS

The real time positioning systems developed by Fugro provide decimeter accuracy by using dual frequency carrier phase observations. These systems and services provide independently high accuracy worldwide positioning using geostationary satellites for correction broadcast.

Recently introduced G2 system is the first system offering combined GPS and GLONASS PPP positioning. Including GLONASS together with GPS satellites improves redundancy, geometry and availability of a positioning solution. It even opens for the possibility to use GLONASS as a positioning system completely independent of GPS also for higher precision positioning adding a new dimension to the redundancy.

REFERENCES

Lapucha D., Barker R., Ott L., Melgard T., Oerpen O. & Zwaan H., 2001. Decimeter-Level Real-Time Carrier Phase Positioning Using Satellite Link. In *Proceedings of The Institute of Navigation GPS 2001 International Technical Meeting, September 11-14, 2001*, Salt Lake City, USA

Melgard T., Vigen E., De Jong K., Lapucha D., Visser H., Orpen O. 2009. G2- The First Real-Time GPS and Glonass Precise Orbit and Clock Service in *Proceedings of The Institute of Navigation GNSS 2009 International Technical Meeting, September 22-25, 2009*, Savannah, USA.

7. Assessing the Limits of eLoran Positioning Accuracy

J. Šafář & F. Vejražka
The Czech Technical University, Prague

P. Williams
The General Lighthouse Authorities of the United Kingdom and Ireland

ABSTRACT: Enhanced Loran (eLoran) is the latest in the longstanding and proven series of low frequency, LOng-RAnge Navigation systems. eLoran evolved from Loran-C in response to the 2001 Volpe Report on GPS vulnerability. The next generation of the Loran systems, eLoran, improves upon Loran-C through enhancements in equipment, transmitted signal, and operating procedures. The improvements allow eLoran to provide better performance and additional services when compared to Loran-C, and enable eLoran to serve as a backup to satellite navigation in many important applications. The Czech Technical University in Prague (CTU) participates in the eLoran research activities coordinated by the General Lighthouse Authorities of the United Kingdom and Ireland (GLAs). In our work we have focused on questions that arise when considering introducing new eLoran stations into an existing network. In particular, this paper explores the issue of Cross-Rate Interference (CRI) among eLoran transmissions and possible ways of its mitigation at the receiver end. An eLoran receiver performance model is presented and validated using an experimental eLoran signal simulator developed by a joint effort of CTU and GLAs. The resulting model is used to evaluate the achievable positioning accuracy of eLoran over the British Isles.

1 INTRODUCTION

In recent years, Global Navigation Satellite Systems (GNSS) have become an integral part of modern society. Be it on land, at sea or in the air, GNSS are an important and often the primary means of Positioning, Navigation and Timing (PNT). Although their qualities make them, in many aspects, superior to other PNT solutions, there is now broad agreement within the radionavigation community that satellite navigation systems are highly vulnerable to unintentional and intentional interference.

The concerns about the vulnerability of GNSS have sparked a renewed interest in the Loran PNT system, or rather in its upgraded version now widely called *enhanced Loran* or simply *eLoran*. The nature of the eLoran system makes its potential failure modes highly independent of GNSS. eLoran is a terrestrial system, which operates in the low-frequency band, uses high-power transmitters and completely different navigation signals. Its signals are also data modulated, which enables eLoran to deliver differential corrections, integrity messages and other data to users. Recently, considerable effort has thus been put into investigating whether eLoran can provide a viable backup to GNSS.

In Europe, the General Lighthouse Authorities of the United Kingdom and Ireland (GLAs) lead the way in eLoran research. The Czech Technical University in Prague (CTU) participates in the eLoran research activities coordinated by the GLAs. In our work we have focused on questions that arise when considering introducing new eLoran stations into an existing network. In particular, this paper explores the issue of *Cross-Rate Interference* (CRI) among eLoran transmissions and its impact on the positioning accuracy performance of eLoran.

In the first part of this paper we give a brief overview of the major factors that determine the achievable positioning accuracy of the system. We then report on the development of an experimental eLoran signal simulator and we demonstrate its use in assessing eLoran receiver performance under noise and interference conditions. Finally, a sample case study is presented that investigates the achievable positioning accuracy of eLoran over the British Isles.

2 ACHIEVABLE POSITIONING ACCURACY OF ELORAN

When referring to accuracy of a positioning system, we need to distinguish between its absolute accuracy and repeatable accuracy. In (USCG COMDTPUB P16562.6), the *absolute accuracy* is defined as the accuracy of a position with respect to the geographic or geodetic coordinates of the Earth. The *repeatable accuracy*, then, is the accuracy with which a user can return to a position whose coordinates have been measured at a previous time with the same navigational system.

Due to the nature of low-frequency signal propagation, Loran systems may suffer from large measurement biases, resulting in absolute accuracy on the order of hundreds of meters. However, Loran's repeatable accuracy is comparable to that of single-frequency (L1) GPS. In the following we briefly discuss the major factors affecting the accuracy performance of eLoran and we explain how eLoran's absolute accuracy can be enhanced to the level of its repeatable accuracy.

2.1 Factors affecting accuracy

Unlike its predecessors, eLoran is a ranging system, which means that obtaining an accurate (2D) position fix generally requires:

1 Accurate Time-of-Arrival (ToA) measurements of signals from at least three transmitters,
2 Accurate ToA to range conversion,
3 Good geometry of the transmitters in view.

Transmitter geometry is a crucial factor in eLoran; however, the impact of geometry on the accuracy performance of a ranging system is well understood and will not be discussed in this paper.

Accurate conversion of ToAs to ranges from transmitters is hampered mainly by signal propagation irregularities when the signals travel over land. In eLoran we account for these irregularities by so-called *Additional Secondary Factors* (ASF). In order to achieve the best possible positioning accuracy, these correction factors in the area of interest need to be measured and stored in the receiver. Fluctuations in the ASF values should also be monitored and broadcast to the user in the form of differential corrections, e.g. using the eLoran data channel.

Table 1. Meeting the maritime accuracy requirement.

Accuracy Limiting Factor	Mitigation
Poor geometry	Installation of additional eLoran transmitters, perhaps using low power mini-eLoran stations as coverage gap fillers
ASF spatial variation	Detailed ASF maps stored in receivers
ASF temporal variation	Differential reference stations generating real-time corrections, broadcast to users e.g. by the eLoran data channel
Uncorrelated noise	Integration time ~ 5 sec is acceptable
Man-made noise and interference	Careful receiver antenna installation, advanced receiver signal processing

Accuracy of the ToA measurements themselves is a function of many variables. It is predominantly determined by the Signal-to-Noise Ratio (SNR) of the received signals. In the Loran frequency band, the dominant sources of noise are atmospheric noise, which is caused by lightning discharges, and man-made noise and local interference from, for example, switch-mode power supplies. Other sources of noise may include transmitter pulse timing jitter or receiver related noise.

Besides noise, another important source of ToA measurement error is interference caused by other radio signals. Currently the biggest source of interference to eLoran is eLoran itself, in the form of CRI. So what exactly is the cause of this interference?

eLoran transmitters are organised in groups of usually 3 to 5 stations called "chains" or "rates". The stations periodically broadcast groups of 8 or 9 specially shaped low-frequency, high-power, pulses (see Figures 2, 3). The interval between successive repetitions of the groups of pulses is unique to each chain and known as the *Group Repetition Interval* (GRI). Careful selection of GRIs and transmission times ensures that stations operating in a chain do not interfere with each other. However, the nature of the system is such that the signals from different chains overlap from time to time (see Figure 4) and may introduce errors into our ToA measurements – this is referred to as CRI.

Another effect of CRI is *transmitter dual-rate blanking*. As a legacy from the Loran-C era, some Loran transmitters are dual-rated, i.e. they broadcast signals on two GRIs. Such transmitters are periodically faced with the impossible requirement of radiating overlapping pulse groups simultaneously. During the time of overlap, those pulses of one group that overlap any part of the other group's blanking interval are suppressed (note e.g. the fourth group of pulses in Figure 2). The *blanking interval* extends from 900 μsec before the first pulse to 1600 μsec after the last.

2.2 Maritime eLoran

Accuracy is the major factor affecting the suitability of eLoran for maritime navigation. IMO standards for the region of Port Approach specify a stringent accuracy requirement of 10 meters (95 percent of the time). A number of studies in the past have shown that accuracies better than 10 m are achievable (Basker et al. 2008, Johnson et al. 2007). Table 1 summarises measures that need to be taken in order to meet the 10 m accuracy requirement in the maritime environment.

From the above it follows that the major error sources in maritime eLoran are the residues of atmospheric noise, transmitter related noise, and CRI. While the impact of the first two factors is well understood and can be modelled (Safar et al. 2010), the issue of CRI has not been sufficiently described so far. In the rest of this paper we will therefore attempt to quantify the effects of CRI and provide CRI models for use in eLoran coverage prediction tools.

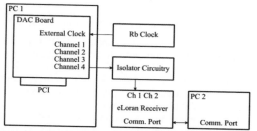

Figure 1. Schematic diagram of the GLA-CTU experimental eLoran signal simulator set-up.

3 ELORAN SIGNAL SIMULATOR

In order to meet the stringent eLoran accuracy performance standards, it is necessary that eLoran receivers employ special CRI mitigation algorithms (Safar et al. 2009). Quantifying the negative effects of CRI is therefore largely a receiver-oriented problem. Unfortunately (but not surprisingly), receiver manufacturers have not widely published the intricacies of their eLoran receivers. In order to get a better understanding of the performance of typical commercial eLoran receivers, an experimental eLoran signal simulator set-up is being developed through cooperation between the GLAs and CTU Prague. Using this set-up, it is possible to work with a receiver in a controlled environment and separate the negative effects of various error sources.

Figure 1 depicts the current simulator set-up. At the heart of the simulator is a DA converter board equipped with four 14-bit converters, providing us with four independent output channels each with a maximum analogue bandwidth of 52.5 MHz. The board is installed in a PC workstation (PC1) and communicates with the host system through the standard 32-bit PCI bus. In the current set-up a stable external 10 MHz clock signal from a GPS-disciplined Rubidium clock is supplied to the board.

The output of the board is connected to the antenna input of the receiver under test through a coupler, which galvanically isolates the receiver's input from the simulator and protects it from overloading. eLoran receivers can either use an E-field "whip" antenna or an H-field antenna. The latter typically consists of two loops whose outputs are combined in the receiver in software in order to provide a beam-steering capability. The simulator currently operates in the E-field (single-channel) mode only. The outputs of the receiver under test are monitored using a separate PC.

The simulator software currently allows the generation of ground wave and sky wave E-field signals, atmospheric noise, and simulation of the pulse timing jitter and transmitter dual-rate blanking. The parameters of the signals are either user defined or calculated for a specified location from corresponding propagation and noise models (Safar et al. 2010). In mathematical terms, the output signal of the simulator can be described as follows:

$$r(t) = \sum_{k=0}^{K-1} \sum_{m=0}^{M-1} \sum_{c=-\infty}^{\infty} \sum_{j=0}^{7} a_{mk} l(t - \tau_{mk} - jT_p - c \cdot T_{GRI,k}) \cdot$$

$$\cos(\omega_0 t + \theta_{mk} + C_{kcj}) + n(t). \qquad (1)$$

Here,

K is the number of eLoran stations "in view" and $T_{GRI,k}$ are their respective group repetition intervals (in seconds);

$M - 1$ is the number of sky waves considered, each with a different amplitude a_m, delay τ_m and phase θ_m ($m = 0$ represents the ground wave);

C_{kcj} are the phase code values (0 or π, according to a standardised pattern), k is the transmitter number, c is the GRI number and j denotes the pulse number within a GRI;

T_p $T_p = 1$ ms;

ω_0 $\omega_0 = 2\pi \cdot 100 \cdot 10^3$ rad/s corresponds to the eLoran carrier frequency of 100 kHz; note, that ω_0 is common to all stations;

$l(t)$ represents the envelope of a single eLoran pulse; for $0 \le t \le 300$ μs it is given by Equation 2 and $l(t) = 0$ otherwise; t_p is the instant when the pulse reaches its maximum value, $t_p = 65$ μs;

$n(t)$ is the noise waveform;

$$l(t) = \left(\frac{t}{t_p}\right)^2 \cdot \exp\left(2 - 2\frac{t}{t_p}\right). \qquad (2)$$

Figures 2-4 show some example eLoran signal waveforms.

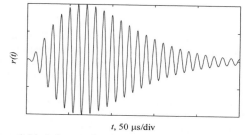

Figure 2. Ideal eLoran pulse (far E-field).

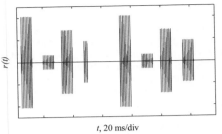

t, 20 ms/div

Figure 3. Simulated ground wave signals of GRI 6731 as would be received at Harwich, UK.

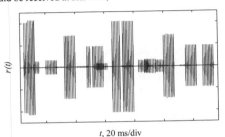

t, 20 ms/div

Figure 4. Simulated ground wave signals of all European chains as would be received at Harwich, UK.

There are no limitations to the number of chains or stations used in the simulation. The simulator therefore provides an excellent tool for studying the effects of CRI.

4 RECEIVER PERFORMANCE MODEL

As discussed earlier, positioning performance of a marine eLoran receiver is primarily determined by the errors in signal ToA measurements. In maritime eLoran, measurement biases are nearly perfectly eliminated through the use of ASF maps and differential corrections. In the following we therefore need be concerned only with the random fluctuations of the ToA error caused mainly by atmospheric noise and CRI, and we will use the standard deviation of the ToA measurements as our performance metric.

4.1 *Basics of eLoran receiver signal processing*

How does an eLoran receiver obtain a ToA measurement at the first place? The ToAs are measured in two stages. First, coarse signal delay relative to the origin of the receiver's time base is measured, based on the shape of the leading edge of the ground wave eLoran pulse. In the model of received signal represented by Equation 1 above, this delay is denoted τ_{0k}. When the approximate ToA is known, carrier phase of the eLoran ground wave signals, θ_{0k}, is measured which allows the receiver to calculate more accurate

ToA values. The coarse estimates are only needed to resolve the ambiguity of the phase measurements; it is therefore the *carrier phase measurement error* which determines the accuracy of our ToAs, and which will be of interest in the following.

4.2 *Receiver performance in white Gaussian noise*

Let us first investigate the impact of atmospheric noise on our measurements. In the first approximation, atmospheric noise may be regarded as a white Gaussian stochastic process. We are therefore facing a problem of estimating the phase of a sinusoid embedded in White Gaussian Noise (WGN). This is a classical problem in estimation theory, and performance analyses of practical phase estimators typically reveal (see e.g. Hua & Pooi 2006) that the variance of the estimates is inversely proportionate to the SNR and the number of signal observations available. In case that the useful signal is a pure sinusoid, the SNR is simply defined as the ratio of the power of the sinusoid to the power of the noise in the signal samples. But how shall we define SNR of an eLoran pulse train?

4.2.1 *Defining SNR*

Unfortunately there is no universally accepted definition of SNR in eLoran; we propose a working definition to be used within this paper.

With conventional Loran signal processing the receiver uses, in the phase estimation process, one signal sample per each received pulse. Signal power can then be defined as the power of a sinusoid having the same amplitude as the envelope of the Loran pulse at the sampling point. There is a hitch, however. The position of the sampling point within the pulse is a compromise between a low SNR at the beginning of the pulse and an increased probability of sky wave contamination later in the pulse; the position is dependent on the receiver's architecture and is generally unknown. Also, the pulse shape is distorted during propagation, reception and signal preprocessing at the receiver, which makes it even harder to determine the effective signal level at the sampling point.

To avoid possible ambiguities, we decided to define SNR external to the receiver. In our simulator experiments we are using the following definition: *SNR is calculated as the ratio of the power of the useful signal at the output of the simulator to the power of the radio-frequency noise present after filtering by the standard front-end filter (8th order Butterworth, 3 dB bandwidth of 28 kHz, centred at 100 kHz).*

This definition assumes the use of ideal signal waveforms (see Equations 1, 2 above) and the power of the useful signal is calculated as the power of a sinusoid having the same amplitude as the ideal eLoran pulse envelope 30 μs into the pulse.

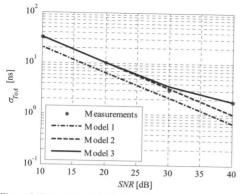

Figure 5. ToA standard deviation vs. SNR for eLoran signals in WGN.

In the performance analysis of a specific receiver we then need to bear in mind that the SNR seen by the receiver's phase estimation algorithms may differ from that above, e.g. due to signal distortion caused by the front-end filter.

4.2.2 Developing the performance model

Based on the cursory analysis above, we may assume that the ToA error model takes the form:

$$\sigma_{ToA}^2 = \frac{c_0 \cdot c_1}{N \cdot SNR} , \qquad (3)$$

where N is the number of signal samples used in the phase estimation process, SNR is expressed as a power ratio as defined above, $c_0 = (1.1254 \ 10^{-6})^2$ takes account of the conversion from phase variance to ToA variance, and c_1 accounts for the pulse distortion during signal pre-processing. Estimated value of this constant for the Reelektronika LORADD receiver used in our analysis, based on information available to the authors, is $c_1 = (1.44)^2$.

Figure 5 shows the predicted ToA standard deviation as a function of SNR. Predictions according to Equation 3 are shown by the dash-dot line (Model 1). In this example it is assumed that the receiver is tracking a GRI 6731 signal and uses a 5 second averaging time, which gives N = 594 pulse samples per ToA measurement.

Also shown in Figure 5 are results of a simulator experiment conducted using the LORADD receiver and our prototype signal simulator (see also APPENDIX A). The actual ToA measurement errors turned out to be a little higher than our predictions. The offset can be calibrated out using another multiplicative constant, $c_2 = (1.55)^2$. The cause of this offset is unclear. The calibrated function is plotted as the dashed line in Figure 5 (Model 2).

It can also be seen from our measurements that the ToA vs. SNR characteristics flattens at high SNRs. This is presumably a result of the receiver's internal noise. The effect can be modelled using an additive constant, $c_3 = (1.5 \cdot 10^{-9})^2$ (solid blue line, Model 3). With the LORADD receiver, however, this effect occurs at very high SNRs unlikely to be encountered in practice, and can safely be neglected.

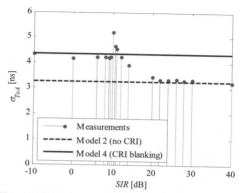

Figure 6. ToA standard deviation vs. SIR; GRI 6731 signal at SNR = 30 dB interfered with signals of GRI 7001 (M,X,Y). All the interfering signals in a particular experiment were set to the same level.

4.3 Receiver performance under CRI conditions

As mentioned before, in order to meet the stringent eLoran performance standards, the impact of CRI within the system must be greatly reduced. Several strategies concerning how the receiver can reduce the effects of CRI have been described in the literature (Pelgrum 2005). There are two prevalent CRI mitigation techniques, commonly referred to as CRI cancelling and CRI blanking.

eLoran employs all-in-view receivers capable of simultaneously tracking signals of many rates. When an eLoran signal is being tracked, a footprint of the received pulse waveform is available. With cancelling, the receiver uses this footprint to reconstruct accurate replicas of the individual signals and suppress the signals of all unwanted rates (Estimate & Subtract). This allows the receiver to mitigate the effects of CRI almost perfectly, however the technique has its limitations, as will be shown shortly.

With CRI blanking, the receiver detects the pulses likely corrupted by CRI and discards them. The interference is thus completely suppressed, but the price we pay is a (sometimes excessive) loss of tracking energy.

4.3.1 Simulator experiments

In order to assess the effects of CRI on a modern eLoran receiver, a series of simulator experiments were conducted in which signals of a selected chain were disturbed by white Gaussian noise and interfered with signals of another chain at different levels. Figure 6 plots the ToA standard deviation versus the Signal-to-Interference Ratio (SIR) for a GRI 6731 signal at 30 dB SNR, interfered with the signals of GRI 7001.

We can see from the plot that for high enough SIR values, the errors are largely determined by the Gaussian noise (see the dashed line in Figure 6, Model 2) and can easily be modelled as described in the previous subsection.

As the interference grows stronger, the measurement errors gradually increase. This gradual increase suggests that in the region of relatively weak interference (SIR above 10 dB) the receiver is using some kind of cancelling algorithm to mitigate CRI. Since the signal replicas used in the CRI cancelling process are mere estimates of the true interfering waveforms, there is always some residual effect on our ToA measurements. This effect is more pronounced as the SIR decreases. With SIR values approaching 10 dB the residual error rises sharply and when the SIR is further decreased, the receiver apparently switches to CRI blanking. A model for the transitional region is currently being developed and will be presented in a follow-up paper. We will now concentrate solely on the CRI blanking.

4.3.2 Modelling the impact of CRI blanking

As explained above, with CRI blanking all the colliding pulses are completely removed from the signal processing. The task of quantifying the impact on the ToA measurements thus reduces to estimating the percentage of discarded pulses and decreasing accordingly the number of samples per ToA measurement in Models 1 to 3 above.

In the following considerations we will ignore the influence of the ninth master Loran pulse, as well as any data modulation of the signals. We will assume that the receiver uses the same blanking strategy as is used on Loran dual-rated transmitters, i.e. that it discards all pulses that overlap any part of the blanking interval of the cross-rating pulse groups (see Figure 7). This is a different approach from the one in our previous paper (Safar et al. 2010), where we had assumed that blanking only occurred when individual pulses overlap each other.

Let us first consider the case of two interfering eLoran ground wave signals. It can easily be shown (Safar et al. 2009) that the average portion of blanked pulses of the desired signal, or the *blanking loss*, can be calculated as:

$$L_b = \frac{w_d + w_i}{T_{GRI,i}},\qquad(4)$$

where w_d is the pulse width for the desired signal, w_i is the width of the blanking interval for the interfering signal, and $T_{GRI,i}$ is the length of the group repetition interval of the interfering station. In our analyses we set $w_d = 250$ µsec, and $w_i = 9500$ µsec.

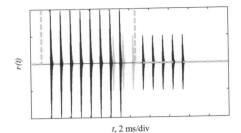

t, 2 ms/div

Figure 7. CRI blanking. Dashed line shows the blanking interval extending over the pulse group of the unwanted cross-rating signal. In this example, samples of the first three pulses of the second group will be discarded.

When analysing real-world eLoran systems we also need to evaluate the *blanking loss due to multiple cross-rating stations*, $L_{b,rx}$. In this case, the evaluation needs to be broken down into two stages. First, we calculate the blanking loss due to stations of individual GRIs, $L_{b,rx,gri}$, by summing the contributions of individual stations, operating on a given GRI. In the following, g denotes the GRI of the interfering station, s identifies individual stations in view, and S_g is the set of stations operating on GRI g:

$$L_{b,rx,gri}[g] = \sum_{s\in S_g} L_{b,rx,st}[s].\qquad(5)$$

Simply summing the blanking loss values is justified, as signals of multiple interferers from a common chain cannot overlap.

Second, we assume that the effects of interference from stations operating on different GRIs are statistically independent, which allows us to calculate the resulting blanking loss as:

$$L_{b,rx} = 1 - \prod_g \left(1 - L_{b,rx,gri}[g]\right).\qquad(6)$$

In addition to ground wave, the effect of *sky wave borne CRI* also needs to be taken into account. The presence of sky waves increases the probability of collision between the interfering pulse trains, depending on the sky wave delay. We model this effect by increasing the width of the blanking interval w_i by the estimated sky wave delay at the point of signal reception.

As mentioned earlier, there is also a loss of signal due to *dual-rate blanking*. In Europe, dual-rated transmitters use priority blanking, where the same rate is always blanked at every overlap (the priority rate is not affected). The loss due to transmitter blanking, $L_{b,tx}$, can then be easily calculated using Equation 4.

The *total blanking loss* for a particular signal of interest, including the effects of dual rate transmitter blanking, $L_{b,tot}$, can be found in a similar fashion as above:

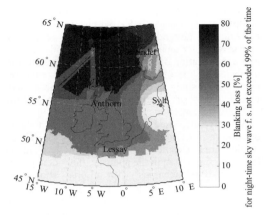

Figure 8. Blanking loss for the 6731 Sylt rate under worst-case sky wave conditions expected.

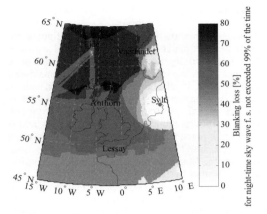

Figure 9. Blanking loss for the 7499 Sylt rate under worst-case sky wave conditions expected.

$$L_{b,tot} = 1 - \left(1 - L_{b,tx}\right)\cdot\left(1 - L_{b,rx}\right). \tag{7}$$

The impact of CRI blanking on the ToA measurement error for a signal of a particular station received at a given SNR can then be estimated using one of the models presented in Subsection 4.2.2, where the number of averaged pulses, N, needs to be reduced accordingly, i.e. we use $(1 - L_{b,tot})\cdot N$ instead of N. With this modification to the model, we achieve nearly perfect agreement with our measurements (see Figure 6, Model 4).

5 CASE STUDY

We will now demonstrate the use of the models presented in this paper through a case study investigating the achievable positioning accuracy of eLoran over the British Isles. Our transmission network will be formed by the 14 transmissions from the 9 European transmitters currently in operation, configured according to Appendix B.

In our study we have made use of the GLA coverage and performance model (Safar et al. 2010) which provides estimates of ground wave and sky wave signal parameters and atmospheric noise values over the area of interest. In accordance with common practice (Last et al. 1991), we have used annual atmospheric noise not exceeded 95% of the time, and night-time sky wave field strength values at 99 percentile, providing a conservative estimate of own sky wave interference. The height of the ionosphere has been assumed to be 91 km.

ToA measurement errors for individual stations have been estimated using the receiver performance models described in Section 4. The signal averaging time has been assumed to be 5 seconds – a typical value for a marine receiver. The receiver is assumed to acquire and track a signal of a particular station only if the SNR is higher than 0 dB (BS EN 61075:1993) and the sky wave field strength to ground wave field strength ratio and sky wave delay are within the limits prescribed by the receiver Minimum Performance Standard (BS EN 61075:1993). CRI at SIR values higher than 10 dB has been assumed to be perfectly cancelled (Model 2); interfering signals at SIR lower than 10 dB and SNR above 0 dB have been blanked (Model 4). *SIR* in our CRI analysis has been defined as the ratio of the power of the ground wave of the useful signal to the power of the interfering signal, calculated either from the ground wave or the sky wave field strength (whichever is higher). As an example of the expected effects of CRI, Figures 8, 9 show the estimated blanking loss for both rates of the dual-rated transmitter at Sylt.

Finally, based on the predicted ToA measurement errors and transmitter geometry, positioning errors have been estimated as described in our previous paper (Safar et al. 2010). Figure 10 shows the predicted 95 percent radius (R95) accuracy calculated under the assumption of Gaussian-distributed measurement errors. As explained above, the plot also assumes that differential eLoran and ASFs are available over the entire area.

6 KNOWN ISSUES & FUTURE WORK

There are a number of reasons why the figures presented in this paper should be interpreted with caution. Let us briefly mention the most important ones.

First of all, we still do not have a rigorous definition of SNR in eLoran. It is therefore difficult to compare measurements obtained using different receivers, and also to translate SNR values from coverage prediction models to actual SNRs as would be seen by a practical receiver. This issue is currently being discussed within the Radio Technical Commission for Maritime services - Special Committee 127 on eLoran Systems (RTCM SC-127).

In developing our receiver performance model we have approximated atmospheric noise by Gaussian-distributed noise. It is well known that real atmospheric noise also contains an impulsive component. eLoran receivers, if properly designed, can benefit from that and may achieve substantial processing gain by suppressing the impulsive part of the noise. In real atmospheric noise conditions, the receiver may therefore perform better than our model predicts. Quantifying the achievable processing gain, however, requires knowledge of the amplitude distribution of the noise (Boyce 2007).

Further performance improvements may be achieved through sky wave aided tracking. Simulator experiments could be conducted to verify this. We might also want to explore alternative sky wave propagation models, such as the USCG-Decca model (Last et al. 1991) which was specifically designed for the Loran frequency band.

On the other hand, there are a number of factors that haven't been considered and may negatively impact the tracking performance. These are for example residual errors due to CRI cancelling, background CRI from distant stations that cannot be tracked, residual Carrier-Wave Interference, or the impact of transmitter timing jitter. These factors may be important at high SNRs.

Finally, we might also want to include differential eLoran in the model. This requires a study of spatial decorrelation of the differential corrections as the user receiver moves away from the reference station. Also the accuracy of ASF maps used in user receivers needs to be assessed and included into the overall error budget.

7 CONCLUSIONS

We have studied the tracking performance of a typical commercially available eLoran receiver under Gaussian noise and CRI conditions. Based on our findings we have developed an updated receiver performance model for the purpose of coverage prediction and optimisation.

Using this new model we have analysed the possibility of mitigating CRI within the European transmission network through blanking at the receiver end. Our analysis suggests that with the current configuration of the network, blanking results in a substantial loss of tracking energy, and we recommend that a study is conducted to examine the potential gains of redesigning the timing of the (e)Loran transmissions in Europe.

We have also used the updated receiver model to generate a positioning accuracy plot for the GLAs' service area. Despite the relatively high blanking loss values assumed in the analysis, the plot suggests that sub-10 m accuracy with eLoran should be achievable in areas of good transmitter geometry, such as off the north and east coast of Britain. *The performance figures presented herein should, however, be interpreted with caution, as this is still work in progress.*

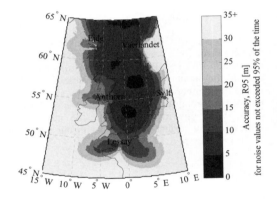

Figure 10. Achievable positioning accuracy of eLoran (R95) under worst-case sky wave conditions expected.

ACKNOWLEDGEMENTS

This work has been supported by the General Lighthouse Authorities of the United Kingdom and Ireland.

REFERENCES

Basker, S. et al. 2008. Enhanced Loran: real-time maritime trials. In *Proceedings of Position, Location and Navigation Symposium, 2008 IEEE/ION.*

Boyce, C.O.L. 2007. *Atmospheric noise mitigation for LORAN.* PhD thesis, Stanford University.

BS EN 61075:1993. *Loran-C receivers for ships - Minimum performance standards - Methods of testing and required test results*, British Standards Institution.

Hua, F. & Pooi Y. K. 2006. ML estimation of the frequency and phase in noise. In *Global Telecommunications Conference, 2006. GLOBECOM '06. IEEE.*

Johnson, G. et al. 2007. Navigating harbors at high accuracy without GPS: eLoran proof-of-concept on the Thames river. In *Proceedings of ION National Technical Meeting*, San Diego, CA, 22-24 January, 2007.

Last, D. et al. 1991. Ionospheric propagation & Loran-C range - the sky's the limit. In *Proceedings of the 20th Annual Technical Symposium, Wild Goose Association*, Williamsburg, VA, 1-3 October, 1991.

Pelgrum, W. 2005. Noise - from a receiver perspective. In *Proceedings of the 34th Annual Convention and Technical Symposium of the International Loran Association*.

Safar, J. et al. 2010. Accuracy performance of eLoran for maritime applications. *Annual of Navigation*, 16:109–122.

Safar, J. et al. 2009. Cross-rate interference and implications for core eLoran service provision. In *Proceedings of the International Loran Association 38th Annual Meeting*, Portland ME.

US Coast Guard COMDTPUB P16562.6 1992. *Loran-C user handbook*.

APPENDIX A   NOTES ON MEASUREMENTS

In our experiments we have been using the Reelektronika LORADD receiver updated with a new firmware developed by Plutargus (v. 1.0), running in the E-field mode.

The LORADD receiver is not capable of measuring absolute ToAs, as it is not equipped with an atomic clock. Instead, we can measure Time Differences (TD) between two selected signals and thus remove the common clock drift. Note, however, that the error in the TD measurements is a combined error, composed of errors of both the signals used in that measurement. In our experiments we have compensated for this effect mathematically.

We have used 1000 seconds worth of data to calculate the tracking errors in Figures 5, 6.

APPENDIX B

Table B.1: European Loran stations.

GRI ID and station name	Dual-rate blanking
6731 Lessay	Priority 6731
6731 Soustons	Not dual-rated
6731 Anthorn	Not dual-rated
6731 Sylt	Priority 7499
7001 Bø	Priority 9007
7001 Jan Mayen	Priority 9007
7001 Berlevag	Not dual-rated
7499 Sylt	Priority 7499
7499 Lessay	Priority 6731
7499 Værlandet	Priority 7499
9007 Ejde	Not dual-rated
9007 Jan Mayen	Priority 9007
9007 Bø	Priority 9007
9007 Værlandet	Priority 7499

8. Fuzzy Evidence in Terrestrial Navigation

W. Filipowicz
Gdynia Maritime University, Poland

ABSTRACT: Measurements taken in terrestrial navigation are random values. Mean errors are within certain ranges what means imprecision in their estimation. Measurements taken to different landmarks can be subjectively diversified. Measurements errors affect isolines deflections. The type of the relation: observation error – line of position deflection, depends on isolines gradients. All the mentioned factors contribute to an overall evidence to be considered once vessel's position is being fixed. Traditional approach is limited in its ability of considering mentioned factors while making a fix. In order to include evidence into a calculation scheme one has to engage new ideas and methods. Mathematical Theory of Evidence extended for fuzzy environment proved to be universal platform for wide variety of new solutions in navigation.

1 INTRODUCTION

In his recent papers the author presented application of Mathematical Theory of Evidence (MTE) in navigation. The Theory appeared to be flexible enough to be used for reasoning on the fix. Contrary to the traditional approach, it enables embracing knowledge into calculations. Knowledge regarding position fixing includes: characteristics of random distributions of measurements as well as ambiguity and imprecision in obtained parameters of the distributions. Relation between observations errors and lines of position deflection is also important. Uncertainty can be additionally expressed by subjectively evaluated masses of confidence attributed to each of observations.

New scheme enabling inclusion of knowledge into the fixing process was presented by Filipowicz 2009c. Way of computation of belief and plausibility as well as location vectors grades can be found in other papers by Filipowicz 2009a, 2009b. Location vectors were constructed assuming normal distribution of measurement errors. The latest was rather a result of limitation imposed on the publications. In order to fill up the hiatus empirical distributions are discussed herein.

Those interested in computational complexity of the fixing algorithms and ways of detecting local maxima should refer to another paper Filipowicz 2010a.

This paper is devoted to a new idea in position fixing in terrestrial navigation. Therefore characteristics of measurements errors are discussed, relation between imprecision of the measured values and lines of position or isolines is also presented.

During computation process abnormally high inaccuracy should be detected. In proposed approach the condition results in large mass of inconsistency, which occurs when no zero mass is assigned to empty sets. High inconsistency mass leads to rejection of the fix or undertaking steps towards fix adjustment. Selected position can be evaluated based on the final inconsistency but also on plausibility and belief values. It should be noted that constant errors are of primary importance when quality of the fix is considered. Using methods that remove systematic deflection of a measurement is recommended. Exploiting horizontal angles instead of bearings makes the fixed position independent from constant errors. The latest is a reason that part of the paper is devoted to the horizontal angle isoline.

MTE exploits belief and plausibility measures, it operates on belief structures. Belief structures are subject to combination in order to increase their initial informative context. The structures can be crisp, interval and fuzzy valued. Mainly crisp valued structures were presented and discussed in the author's previous papers. The structures consist of sets of normal location vectors along with crisp masses of confidence attributed to them. Vectors normality can be achieved through transformation procedure called normalization. Approaches known as Dempster and Yager methods are widely used. Advantages and disadvantages of the two proposals are discussed from nautical usage point of view. Being stuck to the original proposals proved to be not adequate while position fixing. For this reason a modified normalization procedure is proposed in this paper.

2 FUZZY EVIDENCE

Crisp valued standard deviation of a measurement is inadequate. In recent navigation books mean error is described as imprecise interval value usually as: $[\pm\sigma^-_d, \pm\sigma^+_d]$. Mean error of a distance measured with radar variable range marker is within the interval of $[\pm1\% \div \pm1.5\%]$. In the same condition mean error of a bearing taken with medium class radar is within $[\pm1° \div \pm2°]$ as presented by Jurdziński 2008 & Gucma 1995. Using fuzzy arithmetic notation it can be written as a quad (-2, -1, 1, 2). The latest means fuzzy value with core of $[-1°, 1°]$ and support of $[-2°, 2°]$, and reflects the statement that the error is within $[\pm1° \div \pm2°]$. Graphic interpretation of the proposition is shown in Figure 1. The scheme engages probability and possibility theory. Observational errors are assumed to follow a normal distribution. Mean error estimates standard deviation (square root of a variance) of the distribution. The picture shows two confidence intervals related to two different distribution functions. A confidence interval is an interval in which a measurement falls within a range with selected probability. It is assumed that the confidence intervals are symmetrically placed around the mean. A confidence interval with probability equal to 0.683, for the Gauss probability density function is the interval $[\alpha - \sigma, \alpha + \sigma]$ where α is a mean and σ is a standard deviation.

Two confidence intervals introduce imprecision that is usually expressed by an interval or fuzzy value that is a synonym of fuzzy set.

Figure 1 shows trapezoid-like membership function that locates adjacent bearings within the defined set. The function returns possibility regarding given x, it attributes x degree of inclusion within the set. For example abscissa: $x = \alpha+0.5$ fully belongs to the given set, contrary to $x = \alpha+1.5$, its inclusion within the set is partial with degree of membership equal to 0.5. Different membership functions intended for nautical application were discussed by the author in his previous paper Filipowicz 2009a.

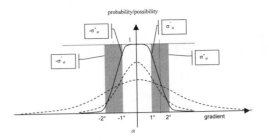

Figure 1. Graphic interpretation of the proposition "bearings mean error is between $\pm1°\div\pm2°$"

Empirical parameters are estimated based on observations. Empirical probability is widely used in practice. In terrestrial navigation it is exploited quite often. Theoretical probabilities are estimated by those calculated from experiments and observations. Empirical probability is the ratio of the number of those results that fall into a selected category to the total number of observations.

The empirical probability estimates statistical probability. Avoiding any assumptions regarding obtained data is the main advantage of estimating probabilities using empirical data. Histograms are widely used as graphical representation of empirical probabilities. Histogram is a diagram of the distribution of experimental data. Usually histogram consists of rectangles, placed over non-overlapping intervals also known as bins. The histogram is normalized and displays relative frequencies. It then shows the proportion of cases that fall into each of several bins. In normalized histogram total area of rectangles equals to one. The bins or intervals are usually chosen to be of the same size. There is no universal rule to calculate number of bins. In the presented application their quantity equals to the number of ranges established around measured value assumed as governed by normal distribution. Empirical distribution of observational errors with imprecise bin width and relative frequencies is shown in Figure 2.

Family of sets $\{\{l_k\}_i\}$ of measured values are given as a result of experiments. Therefore sets of mean values $\{\bar{l}_i\}$ and the bin width s can be obtained. Extreme deflection of means $\Delta\bar{l}^-$ and $\Delta\bar{l}^+$ can be also known. Modal value[1] $\bar{l}_m$ is calculated based on extreme means. Consequently empirical mean and bin widths are interval valued with above mentioned limits. Relative frequencies $\{p_j\}$ for each of considered bins are obtained as crisp or imprecise valued. Formulas from 1 to 4 define complete set of parameters for empirical distributions.

---

[1] Modal value is defined for a fuzzy set. Usually it is calculated as a mean of the set's core, Piegat 2003. It should be noted that modal value is of secondary meaning in distribution characteristics.

$$\left[\Delta \bar{l}^{-}, \ \Delta \bar{l}^{+}\right]=\left[\min_{i}(\{\bar{l}_i\})-\bar{l}_m, \ \max_{i}(\{\bar{l}_i\})-\bar{l}_m\right]$$

where :   (1)

$$\bar{l}_m=\frac{\max_{i}(\{\bar{l}_i\})+\min_{i}(\{\bar{l}_i\})}{2}$$

$$s=\frac{\max_{k,i}(\{\{l_k\}_i\})-\min_{k,i}(\{\{l_k\}_i\})}{n}$$ (2)

$$\left[p_j^{-},p_j^{+}\right]=\left[\min_{k,i}(\{\{p_{jk}\}_i\}),\max_{k,i}(\{\{p_{jk}\}_i\})\right]$$ (3)

$$\left[\Delta s^{-}, \ \Delta s^{+}\right]=\left[\Delta \bar{l}^{-}, \ \Delta \bar{l}^{+}\right]$$ (4)

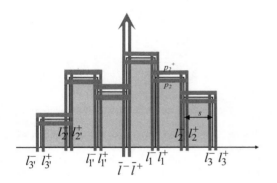

Figure 2. Empirical distribution with imprecise bin width and relative frequencies

3 ISOLINES AND THEIR GRADIENTS

Results of measurements plotted at a chart appear as lines of position. From the mathematic point of view the lines of position are isolines or in many cases lines tangent to them. An isoline for a function of two variables is a curve connecting points where the measurement has the same value. In terrestrial navigation, an isoline joins points of equal bearing, distance or horizontal angel. A bearing is the direction one object is from a vessel. Isoline of a bearing is a line, the same distance from an object produces circle. Isoline of the horizontal angle is also a circle since all inscribed angles that subtend the same arc are equal. The arc joins observed objects. Figure 3 presents isoline of a horizontal angle. A horizontal angle obtained as difference of two bearings is a valuable thing for navigator since it does not contain constant error.

The gradient of a function is a vector which components are the partial derivatives of the function.

For function of two variables gradient is defined by Formula 5.

$$g(x,y)=\nabla f(x,y)=\left[\frac{\partial f}{\partial x}, \frac{\partial f}{\partial y}\right]$$ (5)

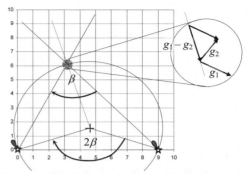

Figure 3. Isoline of a horizontal angle and its gradient in selected point (g_1 and g_2 refers to gradients of the first and second bearing, first bearing is taken to the left object)

Product of the gradient at given point with a vector gives the directional derivative of the function in the direction of the vector. The direction of gradient of the function is always perpendicular to the isoline. Gradients, measurements error and lines of position deflections are dependent values. Formula 6 shows the relation.

$$M(x,y)=\frac{\sigma}{|g(x,y)|}$$ (6)

Table 1. Parameters of the horizontal angle isoline

isoline parameter	formula
isoline radius	$r=\dfrac{d_{12}}{2\sin(\beta)}$
center coordinates	$(x_1+r\cos(90-\beta+\theta), y_1+r\sin(90-\beta+\theta))$
gradient module	$\|g\|=\dfrac{d_{12}}{d_1 d_2}\left[\dfrac{\text{rad}}{\text{Nm}}\right]$

d_{12} – distance between observed objects
θ – inclination, related to x axis, of the line passed through the objects ($\theta=0$ in Figure 3)
x_1, y_1 – coordinates of the left object (see Figure 3)
β – horizontal angle calculated as difference of bearings ($\beta > 0$)
d_i – distance to i-th object ($d_i \neq 0$)

Error of the measurement divided by the module or length of the gradient in selected point gives deflection of the isoline at the point. In the proposed solution limits of introduced strips and possible isolines coverage are to be calculated accordingly.

Radius length, coordinates of the center and gradient module for horizontal angle isoline can be calculated with formulas presented in Table 1.

Table 2 contains data regarding isoline shown in Figure 3. The data embrace distances, gradients modules and isoline errors calculated for measurements standard deviation of ±1°. Appropriate values were obtained for selected points placed in the isoline.

Table 2. Selected points at the horizontal angle isoline, gradients and isoline errors

x	0	1	3	4	5	6	8	9
y	3.2	4.9	6.2	6.4	6.4	6.2	4.9	3.2
d_1 [Nm]	3.2	5.0	6.8	7.5	8.1	8.6	9.4	9.6
d_2 [Nm]	9.6	9.4	8.6	8.1	7.5	6.8	5.0	3.2
$\lvert g \rvert \left[\frac{°}{\text{Nm}}\right]$	16.7	11.1	8.8	8.5	8.5	8.8	11.1	16.7
$\pm M$ [cables]	0.60	0.90	1.14	1.18	1.18	1.14	0.90	0.60

isoline error M was calculated for measurement mean error $\sigma = \pm 1°$

Isoline of a horizontal angle and its limits calculated for interval $[\beta - 3\sigma, \beta + 3\sigma]$ is shown in Figure 4. Limits of an isoline shows its extreme shifts due to measurements errors. These limits can be of the same size along the line as for example for distances. For bearings and horizontal angles limits vary depending on the position of the observer. Within the limits strips related to confidence intervals are established. Levels of confidence and way of selecting stripes are discussed in the paper by Filipowicz 2010b.

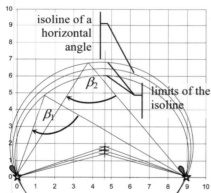
Figure 4. Isoline of the horizontal angle and its limits

4 SCHEME OF A POSITION FIXING

Let us consider three rectangular ranges related to three isolines as shown in Figure 5. Within ranges six strips were distinguished. Widths of the strips are calculated based on measurement errors and the isoline gradients. Each strip has fuzzy borders depending on imprecision in estimations of the isoline errors distribution. Theoretical or empirical probabilities of containing the true isoline within strips are given. Having particular point and all before mentioned evidence support on representing fixed position for given point should be found. This is quite different from traditional approach where single point should be found and available evidence hardly exploited.

The scheme of approach is as follows:
Given: available evidence obtained thanks to nautical knowledge
Question: what is a support that particular point can be considered as fixed position of the ship?

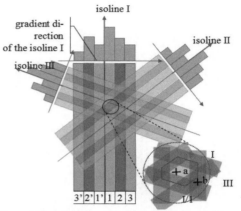
Figure 5. Three isolines with strips established around them

Figure 5 shows common area of intersection of three areas associated with three isolines. Six strips were selected around each isoline, the strips were numbered as shown in the figure. Number 3' refers to the far most section, number 3 indicates closest range according to gradient direction and regarding observed object(s). Assuming normal or empirical distribution probabilities attributed to each of the strips might be as shown in Table 3.

Table 3. Example probability values

strip	3'	2'	1'	1	2	3
normal distribution	0.021	0.136	0.342	0.342	0.136	0.021
empirical distribution	0.05	0.15	0.30	0.35	0.10	0.05

Figure 5 also shows magnified fragment of the area with two points situated within it. Points are marked with a and b. For both points hypothesis that they represent fixed position will be calculated. Support that point a can be considered as a fix is justified by the following probabilities related to (note that point a is entirely situated within crossing strips):
– membership within strip 1 regarding isoline I
– membership within strip 2' regarding isoline II
– location within strip 1' regarding isoline III

Position of point b is partial within strips related to isolines II and III. Its memberships are estimated as follows: II/2'→0.3, II/1'→0.7, III/1'→0.9,

III/2'→0.1. Thus support that point b can be considered as a fix is justified by the following:
- full membership within strip 1 with reference to isoline I
- partial location within strip 2' regarding isoline II
- partial location within strip 1' regarding isoline II
- partial membership within strip 1' with reference to isoline III
- partial location within strip 2' with reference to isoline III

Evaluation of each of the measurements should also be included in calculation. Navigator knows which observation is good or bad, which are preferable to the others. Usually the opinion is subjective and can be expressed as linguistic term or a crisp value.

Table 4. Example probability values

strip	mass of evidence	sets	ref. I		ref. II.		ref. III	
3'	$m_{3'} = 0.05$	$\mu_{i3'}$	0	0	0	0	0	0
2'	$m_{2'} = 0.15$	$\mu_{i2'}$	0	0	1	0.3	0	0.9
1'	$m_{1'} = 0.35$	$\mu_{i1'}$	0	0	0	0.7	1	0.1
1	$m_1 = 0.30$	μ_{i1}	1	0.5	0	0	0	0
2	$m_2 = 0.10$	μ_{i2}	0	0.5	0	0	0	0
3	$m_3 = 0.05$	μ_{i3}	0	0	0	0	0	0
uncertainty			0.3		0.2		0.1	

ref. stands for reference to:
index i indicates isolines (I, II or III)

Table 4 contains preliminary results of the example analysis. The table contains fuzzy points locations within selected strips, locations are given with reference to each of the isolines. Example empirical probabilities are included in column 2. Last row presents uncertainty, weights of doubtfulness, which is a complement of credibility, attributed to each measurement.

Belief structure is a mapping or an assignment of masses to normal location sets. Location vectors are to be normal it means that their highest grade must be one. Subnormal sets should be converted to their normal state using normalization procedure. Vectors are supplemented with all one set, which expresses uncertainty. It says that each location is equally possible. Mass attributed to this vector shows lack of confidence to a particular measurement. Thanks to this value all observations can be subjectively differentiated. All location vectors have assigned mass of confidence. Appropriate values are calculated as a product of empirical probability assigned to particular strip and complement of uncertainty related to given measurement. It should be noted that the sum of all masses within a single belief structure is to be equal to one. Table 5 presents three normalized belief structures constructed based on data from Table 4.

Belief structures are subject of combination in order to obtain knowledge base enabling reasoning on the position of the ship. It is known that combination

of belief structures increase their initial informative context. By taking several distances and/or bearings a navigator is supposed to be confident on true location of the ship.

Plausibility and belief of the proposition represented by a fuzzy vector included in collection of result sets are calculated. In position fixing plausibility is of primary importance, for discussion on this topic see Filipowicz 2009a, 2010c. To calculate final plausibility and belief one has to use formulas presented by Denoeux 2000, the expressions were further simplified by the author Filipowicz 2010c. In presented example plausibility values that given points can be selected as a fixed position are: $pl_a = 0.62$, $pl_b = 0.60$. Obviously a dense mesh of points is to be considered in practical implementations.

Table 5. Final normalized belief structures

b.s. I			b.s.II			b.s. III		
{1	0.5}	0.21	{1	0.3}	0.12	{0	1}	0.08
{0	1}	0.07	{0	1}	0.28	{1	0.1}	0.31
{1	1}	0.72	{1	1}	0.60	{1	1}	0.61

b.s. stands for belief structure

5 NOTES ON NORMALIZATION OF PSEUDO BELIEF STRUCTURES

Two strips that do not embrace the common points are disjunctive and their intersection is empty. Result of combination of the disjunctive vectors is a null set. Therefore product of masses attributed to both combined disjunctive vectors is assigned to empty set what means occurrence of inconsistency. Inconsistency results in a pseudo belief structure that must be converted to its normal state. Two normalization procedures are used: one was proposed by Dempster another one by Yager. At first both of them considered crisp vectors. Further extensions for fuzzy environment were suggested by Yager 1995. Although it is quite often that many authors refer to them using original methods inventor names. Normalization procedures are quite different in two aspects, namely in allocation of inconsistency masses and modification of fuzzy sets contents called grades. Masses of inconsistency in Dempster approach increase weights attributed to not null sets. In Yager proposal the masses increase uncertainty. In case of subnormal sets Dempster suggested division by highest grade. It preserves allocation of points within selected strips. Yager proposed adding complement of the largest grade to all elements of the set. It corrupts allocation of points within selected strips. Therefore results of subnormal belief structures conversion to their normal state using the two methods are different, see Table 6 for case study. Fuzzy sets are location vectors containing fuzzy memberships of a search space points within selected strips. Thus Dempster transformation causes that points with not null locations increase their memberships, empty grades are not changed. In Yager normalization all considered points gain some degrees of membership. Unfortunately it may adversely affect computational process and ability of evaluation of the obtained fix. Therefore modified normalization method is proposed. In the approach inconsistency masses increase uncertainty very much like in Yager method. Conversion of subnormal sets remains in line with Dempster proposal. In order to obtain proper grades all of them are divided by the highest one. Modified method preserves location of search space points. The method also enables identification of all inconsistency cases as depicted by Filipowicz 2010b.

Table 6. Two example fuzzy sets, their normalizations and combinations

	Location vectors									$m(..)$
μ_l	{0	0.8	0	0	0	0	0	0.6	0}	0.41
μ_l^Y	{0.2	1	0.2	0.2	0.2	0.2	0.2	0.8	0.2}	0.41
μ_l^D	{0	1	0	0	0	0	0	0.75	0}	0.48*)
μ_l^M	{0	1	0	0	0	0	0	0.75	0}	0.33
μ_2	{0	0	0	0.67	0	1	0	0	0}	0.20
$\mu_{\mu1}{}^Y{}_{\wedge\mu2}$	{0	0	0	0.2	0	0.2	0	0	0}	0.08
$\mu_{\mu1}{}^D{}_{\wedge\mu2}$	{0	0	0	0	0	0	0	0	0}	0.10
$\mu_{\mu1}{}^M{}_{\wedge\mu2}$	{0	0	0	0	0	0	0	0	0}	0.07

*) - according to Dempster proposal masses of non empty sets are modified during normalization
μ_l^Y - fuzzy set μ_l normalized with Yager method
μ_l^D - fuzzy set μ_l normalized with Dempster method
μ_l^M - fuzzy set μ_l normalized with modified method
$\mu_{\mu1}{}^Y{}_{\wedge\mu2}$ - result of combination of fuzzy sets μ_l^Y and μ_2
$\mu_{\mu1}{}^D{}_{\wedge\mu2}$ - result of combination of fuzzy sets μ_l^D and μ_2
$\mu_{\mu1}{}^M{}_{\wedge\mu2}$ - result of combination of fuzzy sets μ_l^M and μ_2

Table 7. Dempster versus Yager versus modified approaches

	Dempster normalization (Yager smooth normalization)*)	Yager normalization	modified normalization
way of modification of masses assigned to not null sets	increased by a factor calculated using inconsistency values	remain unchanged	reduced by complement of the highest grade
result uncertainty	solely depend on initial uncertainties	uncertainty is increased by total mass of inconsistency	increased by reduction of not null sets masses
modification of membership grades	general image of location vectors is preserved, null grades remain unchanged	null grades of location vectors gain some membership	general image of location vectors is preserved
ability to detect all inconsistency cases	possible	impossible	possible
recommendation	belief structures with fuzzy location vectors	belief structures with binary location vectors	belief structures with fuzzy location vectors
not recommended for	belief structures with binary vectors and high inconsistency	belief structures with fuzzy vectors and high inconsistency	belief structures with binary vectors and high inconsistency
computational complexity	rather high	rather low	rather low
final solution affected by high inconsistency	not observed	might adversely affect final solution	not observed

*) original method name suggested by Yager 1995

Table 6 embraces example of two fuzzy sets that are excerpted from belief structures. First of the sets is subnormal and needs to be converted. Their normal states obtained by three different methods are also presented. Results of combinations of the converted sets with the second one are included in last three rows of the table.

Combination is carried out using minimum operator and product of masses involved. Formula 7 delivers proper expressions.

$$\mu_{\mu_1 \wedge \mu_2}(x_i) = \min(\mu_1(x_i), \mu_2(x_i))$$
$$m(\mu_{\mu_1 \wedge \mu_2}(x_i)) = m(\mu_1(x_i)) \cdot m(\mu_2(x_i))$$
(7)

Masses of credibility assigned to all vectors and to results of their combinations are shown in the last column of Table 6.

Table 7 contains comparison of Dempster, Yager and modified normalizations taking into account practical aspects presented in first column. It should be noted that position fixing engages fuzzy location vectors therefore modified normalization should be recommended. Most important feature of the Dempster and modified methods is ability to preserve general shape of location vectors, null grades remain unchanged. Consequently all inconsistency cases can be detected.

6 SUMMARY AND CONCLUSIONS

Bridge officer has to use different navigational aids in order to refine position of the vessel. To combine various sources he uses his common sense or relies on traditional way of data association. So far Kalman filter proved to be most famous method of data integration. Mathematical Theory of Evidence delivers new ability. It can be used for data combination that results in enrichment of their informative context. The Theory extension to a fuzzy platform proposed by Yen 1990 enables wider and more complex applications.

Based on the Theory concept new method of position fixing in terrestrial navigation is proposed. The method enables reasoning on position fixing based on measured distances and/or bearings. It was assumed that measured values are random ones with theoretical or empirical distribution. Knowledge on used aids and observed objects is included into combination scheme. Relation between measurement error and deflection of the isoline was also depicted. It was suggested that instead of bearings concept of horizontal angles should be used, obtained isoline is constant error free.

The true isoline of distance, bearing or horizontal angle is somewhere in the vicinity of the isoline linked to a measurement. To define true observation location probabilities six ranges were introduced. Probability levels assigned to each strip can be calculated based on features of normal distribution or they can be delivered from experiments. Standard deviation of the distribution is assumed to be within known range. Empirical data also varies within some range. In both cases imprecise interval valued limits of ranges are to be adopted. Sigmoid membership functions are used for establishing points of interest levels of locations within established ranges. Calculated locations are elements of fuzzy sets called location vectors. Vectors supplemented with the one expressing uncertainty compose one part of belief structure. Another part embraces masses of initial believes assigned to location vectors and uncertainty. Complete belief structure is related to each of measurements. Mass assigned to uncertainty expresses subjective assessment of measuring conditions. One has to take into account: radar echo signature, height of objects, visibility and so on to include measurement evaluation. Fuzzy values such as poor, medium or good can be used instead of crisp figures. Imprecise masses values engage different way of calculation and will be discussed in a future paper.

Belief structures are combined. During association process search space points within common intersection region are selected. Result of association is to be explored for reasoning on the fix. All associated items are to be taken into account in order to select final solution.

Mathematical Theory of Evidence requires that mass of evidence assigned to null set is to be zero and fuzzy sets are to be normal. Assignment for which above requirements are not observed is pseudo belief structure and is to be normalized. Pseudo belief structures can occur at the structures prepara-

tion stage as well as during association process. Usually null sets are results of combination of two ranges or areas without common search space points. The occurrences indicate abnormality in computation that might result from extraordinary erroneous measurements and/or wrongly adjusted search space. Therefore all null assignment cases are to be recorded and analyzed. Two normalization procedures proposed by Dempster and Yager are widely used. Converting procedures are quite different in two aspects. Masses of inconsistency in Dempster approach increase weights attributed to not null sets. In Yager proposal the masses increase uncertainty. In case of subnormal sets Dempster suggested division by highest grade, Yager proposed adding complement of the largest grade to all elements of the set. The latter causes that none of these approaches should be perceived as superior in case of position fixing. Therefore modified scheme was proposed. It takes best things from both proposals. Way of conversion of subnormal sets is taken from Dempster method and managing of inconsistency comes from Yager approach.

REFERENCES

Denoeux, T. 2000. Modelling vague beliefs using fuzzy valued belief structures. *Fuzzy Sets and Systems* 116: 167-199.

Filipowicz, W. 2009a. Belief Structures and Their Application in Navigation. *Methods of Applied Informatics* 3: 53-83. Szczecin: Polska Akademia Nauk Komisja Informatyki.

Filipowicz, W. 2009b. Mathematical Theory of Evidence and its Application in Navigation. In Adam Grzech (eds.) *Knowledge Engineering and Expert Systems*: 599-614. Warszawa: Exit.

Filipowicz, W. 2009c. An Application of Mathematical Theory of Evidence in Navigation. In Adam Weintrit (ed.) *Marine Navigation and Safety of Sea Transportation*: 523-531. Rotterdam: Balkema.

Filipowicz, W. 2010a. Fuzzy Reasoning Algorithms for Position Fixing, Pomiary Automatyka Kontrola 56. Warszawa (in printing).

Filipowicz, W. 2010b. Belief Structures in Position Fixing. In Jerzy Mikulski (ed.) *Communications in Computer and Information Science* 104: 434-446. Berlin. Heidelberg: Springer.

Filipowicz, W. 2010c. New Approach towards Position Fixing. *Annual of Navigation* 16: 41-54

Gucma, S. 1995. Foundations of Line of Position Theory and Accuracy in Marine Navigation. Szczecin: WSM.

Jurdziński, M. 2005. Foundations of Marine Navigation. Gdynia: Gdynia Maritime University.

Piegat, A. 2003. Fuzzy Modelling and Control. Warszawa: EX-IT.

Yager, R. 1996. On the Normalization of Fuzzy Belief Structure. *International Journal of Approximate Reasoning*. 14: 127-153.

Yen, J. 1990. Generalizing the Dempster-Shafer theory to fuzzy sets. *IEEE Transactions on Systems, Man and Cybernetics*. 20(3): 559-570.

9. Ground-based, Hyperbolic Radiolocation System with Spread Spectrum Signal - AEGIR

S.J. Ambroziak, R.J. Katulski, J. Sadowski, W. Siwicki & J. Stefański
Gdansk University of Technology, Poland

ABSTRACT: At present the most popular radiolocation system in the world is Global Positioning System (GPS).As it is managed by the Department of Defence of the U.S.A., there is always the risk of the occasional inaccuracies or deliberate insertion of errors, therefore this system can not be used by secret services or armies of countries other than the U.S.A. This situation has engender a need for development of an autonomous, ground-based radiolocation system, based on the hyperbolic system with spread spectrum signals. This article describes the construction and operation of such a system technology demonstrator which was developed at the Technical University of Gdansk. It was named AEGIR (god of the ocean in Norse mythology). This paper presents preliminary results and analysis of its effectiveness.

1 ASSUMPTIONS OF DESIGNED SYSTEM

The starting point is to build a ground-based system which is mainly associated with the hyperbolic localization systems. They are based on the differential measurement method called Time Differential of Arrival (TDOA). The first hyperbolic system (Gee) appeared during Second World War. It has evolved (DECCA, OMEGA), but the moment satellite navigation appeared, they have practically gone out of use. Up to now only LORAN C system is still underlined.

Our goal has been to create a system of hyperbolic localization but made in modern technology. The designed system uses spread spectrum signals. The second element is an asynchronous operation. The system resigns chain relationship between stations. With this approach, our system has gained new features and new functionality compared to traditional solutions.

The first task is to determine the basic parameters, i.e.: frequency, bandwidth, modulation, etc. After careful consideration, the following parameters has been set:

- spread spectrum signals (using DS-CDMA)
- reliance on a hyperbolic system (TDOA method)
- frequency: 431.5 MHz
- the width of the transmission channel – 1 MHz
- transmission speed of navigational information - 1 kb/s.
- modulation: QPSK

2 HYPERBOLIC SYSTEMS – TDOA METHOD

The TDOA method, as mentioned before, is based on a calculation of the time difference between stations. Suppose there are N ground stations, the coordinates for the i-th station are $S_i = (x_{Si}, y_{Si})$, where i = 1, ..., N, and the search object's coordinates are $M = (x_M, y_M)$.

If you define a signal propagation time between the i-th station and the searched position in the point M as T_i, so the distance between the i-th station and the point M is as follow:

$$d_i = T_i \cdot c = \sqrt{(x_{Si} - x_M)^2 + (y_{Si} - y_M)^2}, \qquad (1)$$

where:
c - velocity of wave propagation ($3 * 10^8$ m / s)
T_i - the propagation delay between the i-th station and the point M,
d_i - distance between i-th station and the point M.

Timing differences between the i-th station and a first one, can be written as:

$$T_{i1} = T_i - T_1 \qquad (2)$$

Differences in the distances between those stations, can be described by the following relationship:

$$d_{i1} = T_{i1} \cdot c = d_i - d_1, \qquad (3)$$

After putting equation (1) in equation (3) we obtain hyperbolic equation:

$$d_{i1} = \sqrt{(x_{Si} - x_M)^2 + (y_{Si} - y_M)^2} \atop - \sqrt{(x_{S1} - x_M)^2 + (y_{S1} - y_M)^2}. \quad (4)$$

Equation 4 presents the difference in distance between the first and i-th station.

Determination of the distance difference between another pair of base stations generates more hyperbolas and a point of their intersection gives us a position. There are many algorithms [1-4], which allow to determine the coordinates, however for the purpose of the system the Chan method was chosen [1].

The principle of TDOA method can be illustrated as follows. Assume that we have three reference stations positioned as in Figure 1.

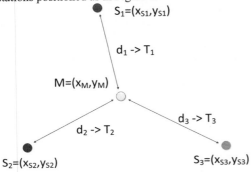

Figure 1. Deployment of ground stations to illustrate the method of TDOA

Propagation time from the station to your desired position in the point M is respectively T_1, T_2 and T_3 and the distance between them is d_1, d_2 and d_3. Each station has coordinates as follows: S1=(x_{S1}, y_{S1}), S2=(x_{S2}, y_{S2}) and S3=(x_{S3}, y_{S3}).

Determination of temporary differences between the stations is illustrated in Figure 2. It has been assumed that each station transmits at the same time an impulse signal. Figure 2a shows the moment of broadcasting signals by the station. Figure 2b shows the time of receipt of the impulses at the point of searched position.

Analyzing Figure 2 it can be observed that when the impulses are transmitted at the same time from each ground station, the time difference at the receiver side is easily measured. Unfortunately, such a synchronization is difficult to obtain.

For this reason, the system has been designed as asynchronous one. This allows switching off and on any station without resynchronization the system. In order to implement this feature, it has been necessary to create a reference station, which not only transmits, but is also able to receive signals from neighbouring stations. With this approach, the reference station measures the time differences in synchronization between the reference signal and its

neighbouring stations so the calculated time differences are sent to the receiver. This mode of operation is illustrated in Figure 3 [8].

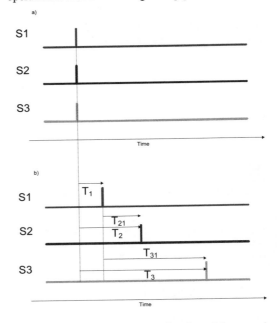

Figure 2. Timing between signals broadcasted by ground stations a) the moment of broadcasting impulses by the stations b) the time of receipt of impulses by the receiver

As in the previous example, stations transmit a reference signal as an impulse, but time of broadcasting these impulses, as shown in Figure 3, is random. The stations have the ability to "listen to" neighbouring stations. This is illustrated in Figure 3b. Reference station designated as S1 receives signal from other two stations: S2 and S3, and calculates the time difference between its own and these stations' signals (nT_{21} and nT_{31}). These time differences are then sent to the receiver. The receiver (pictured in Figure 3c) sets its own time difference between the received impulses from the reference station (dT_{21} and dT_{31}). Additionally, each ground station sends to the receiver its own coordinates (respectively x_{S1}, y_{S1} - the coordinates of the first station, x_{S2}, y_{S2} - coordinates of the second station and x_{S3} i y_{S3} - coordinates of the third station), so that the receiver calculates the propagation time between the reference stations (T_{S1S2}, T_{S1S3}). Taking into account all sent data, the receiver calculates a real difference in propagation time between stations, which present the following equation:

$$T_{21} = nT_{21} - dT_{21} - T_{S1S2} \atop T_{31} = nT_{31} - dT_{31} - T_{S1S3} \quad (5)$$

The time differences defined in this manner allow to determine coordinates of searched object M using one of the sets of algorithms [1-4].

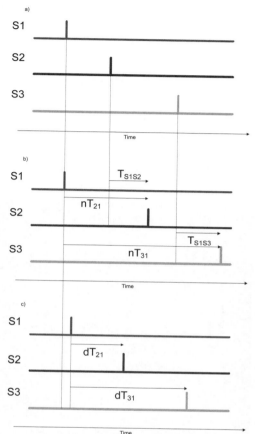

Figure 3. Timing between signals broadcasted by base stations in an asynchronous system, a) the moment of broadcasting impulses by the stations b) the time of receipt of impulses by S1 station c) the time of receipt of impulses by the receiver [8]

In case of reception from only three stations, Chan's algorithm will result in a set of two coordinate values. Only one of them is correct and the other one lies outside the presented area [7].

3 HARDWARE IMPLEMENTATION

The system consists of a localizer/receiver and ground/reference stations.

The block diagram of a receiver is presented in Figure 4.

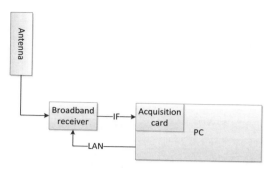

Figure 4. Block diagram of a receiver

The receiver has been made in the technology of Software Defined Radio [5]. It consists of: an antenna, a broadband receiver, an analog to digital converter (in the form of data acquisition card) and digital signal processor (in form of PC). This approach allows to shape flexibly functionality of the receiver. Hardware implementation of a receiver is presented in Figure 5.

Figure 5. Hardware implementation of the receiver

Ground stations, as it was mentioned before, have the ability to "listen to" neighbouring stations. It is assumed that the system should consists only of such stations (Master ones). However for demonstrable purposes only one Master station is required. Therefore two types of ground stations were created: broadcasting stations (Slave type) and broadcasting and listening ones (Master type).

The block diagram of a Slave station is shown in Figure 6.

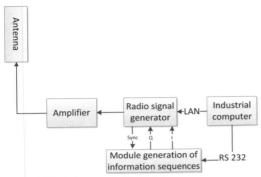

Figure 6. Block diagram of a Slave station

The main element of the station is a radio signal generator, whose task is to broadcast modulated signal with data that are generated by industrial computer. Hardware implementation of a Slave station is shown in Figure 7.

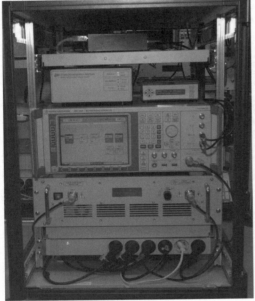

Figure 7. Hardware implementation of a Slave station

The block diagram of the last element of the described system - Master station – is shown in Figure 8.

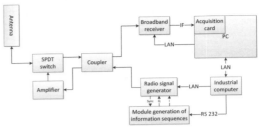

Figure 8. Block diagram of a Master station

Master station is a combination of a receiver and a Slave station. The task of the receiver is to listen to a nearby station and to determine difference in synchronization between reference signal and signals from the neighbouring stations. Hardware implementation of a Master station is shown in Figure 9.

Figure 9. Hardware implementation of a Master station

As already mentioned, the system uses spread spectrum signals. Broadcasted signals are called Navigation Messages and they are divided into two types. First type – called Basic Navigation Message (BNM), contains information of geographic coordinates of reference station, the height of the suspension of the antenna, transmitter power, etc. The second one – Additional Navigation Message (ANM) – contains previously mentioned time differences between stations.

4 TESTS AND RESULTS

The developed technology demonstrator was tested twice in real conditions. The area of our tests was the Bay of Gdansk. The first measurements were carried out in April 2010. with the three ground stations located: first - on top of the CTM building, second – on the top of the lighthouse in Hel and third – on the top of the lighthouse in New Port. The receiver was placed on a small watercraft. Those tests allowed us to find the underdeveloped parts of the system and suggested new approaches. Subsequent measurements were carried out six months later, in October 2010. For the purpose of these tests, the fourth reference station (Slave one) was added; it was installed on the top of the building of the Faculty of Electronics, Telecommunications and Informatics of Gdansk University of Technology (positions of all four ground stations are presented in Figure 10).

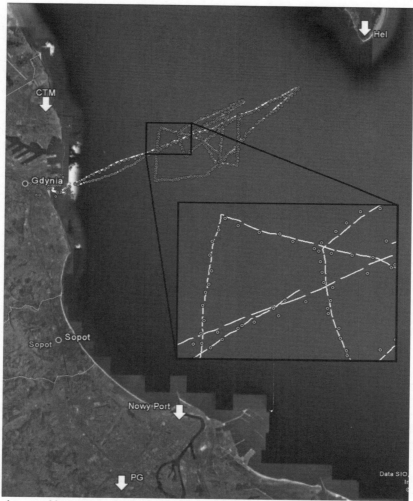

Figure 10. Deployment of four ground stations (arrows) and a path of GPS and GLONASS positions (doted line) and readings of autonomous AEGIR system (dots)

During field tests a position from a satellite navigation system was recorded with the use of a Javad Alpha receiver, which enables simultaneous reception from both American (GPS) and Russian (GLONASS) systems.

The effects of our tests are illustrated by the visualization shown in Figure 10, created with use of Google Earth software. The dotted line represents the path of positions received from the satellite systems GPS/GLONASS, and the dots represent the calculated positions of the ground-based system. Analyzing the visualization shown in Figure 10 it can be observed how accurate the route travelled by the vessel was reconstructed by points calculated by the autonomous localization system - AEGIR.

5 CONCLUSIONS

The presented results are the preliminary approach to the analysis of a designed system. The first results suggest that this solution can be very useful for military purposes.

The presented system has been developed to be very flexible. It allows to use more than three ground-stations. Placing them in areas of known positions, allows to create a grid, which will provide an independent reading of coordinates from satellite systems.

The presented system is fully asynchronous. In case of damage or shutdown of one of the stations, the system in a short time will be again fully functional. The only condition is to receive signals from at least three ground stations.

ACKNOWLEDGEMENT

The described research is funded by the Polish Ministry of Science and Higher Education as a part of research and development project No O R00 0049 06. The authors express their sincere thanks for allocated funds for this purpose.

REFERENCES

[1] Chan Y.T., Ho K.C., *"A Simple and Efficient Estimator for Hiperbolic Location"*, IEEE Transactions on signal processing, pp. 1905-1915, vol. 42, no. 8, August 1994,
[2] Foy W.H., *"Position-Location Solutions by Taylor-Series Estimation"*, IEEE Trans. On Aero. And Elec. Systems, vol. AES-12, no. 2, 1976, pp. 187-194.
[3] Fang B.T., *"Simple Solution for Hiperbolic and Related Position Fixes"*, IEEE Trans. On Aero. And Elec. Systems, vol 26, no. 5, 1990, pp. 748-753.
[4] Friedlander, *"A Passive Localization Algorithm and Its Accuracy Analysis"*, IEEE Jour. Of Oceanic Engineer., vol. OE-12, no. 1, 1987, pp. 234-245.
[5] Katulski R., Marczak. A., and Stefański J.; *"Software Radio Technology"* (in polish), Telecommunication review and telecommunication news No. 10/2004, pp. 402-406.
[6] Ambroziak S.J., Katulski R.J., Sadowski J., Siwicki W., Stefański J., *"Autonomous ground-based and self-organized radiolocation systems"* (in polish), Naval Armament & Technology Conference NATCon'2010, Gdynia 20-21.X.2010.
[7] Stefański J. *"Methods' analysis for position estimation of movable terminal in a multipath environment"* (in polish), KKRRiT 2009, Telecommunication review and telecommunication news No. 6/2009, pp.364-367.
[8] Patent application no. P393181, "Asynchronous system and method for determining position of persons and/or objects" (in polish), 2010-12-08.

10. An Algorithmic Study on Positioning and Directional System by Free Gyros

T.-G. Jeong & S.-C. Park
Korea Maritime University, Busan

ABSTRACT: The authors aim to establish the theory necessary for developing free gyro positioning system and focus on measuring the nadir angle by using the motion rate of a free gyro. The azimuth of a gyro vector from the North can be given by using the property of the free gyro. The motion rate of the spin axis in the gyro frame is transformed into the platform frame and again into the NED (north-east-down) navigation frame. The nadir angle of a gyro vector is obtained by using the North components of the motion rate of the spin axis in the NED frame. The component has to be transformed into the horizontal component of the gyro by using the azimuth of the gyro vector and then has to be integrated over the sampling interval. Meanwhile the authors suggest north-finding principle by the angular velocity of the earth's rotation. That is, ship's heading is obtained by using the fore-and-aft and athwartship components of the motion rate of the spin axis in the NED frame.

1 INTRODUCTION

A free gyro positioning system (FPS), which determines the position of a vehicle by using two free gyros, was first suggested by Park & Jeong(2004). It is originally an active positioning system like an inertial navigation system (INS) in view of obtaining a position without external source. However, a FPS is to determine its own position by using the angle between the vertical axis of local geodetic frame and the axis of free gyro (hereinafter called 'nadir angle'), while an INS is to do so by measuring its acceleration.

The errors in the FPS were investigated broadly by Jeong(2005). And the algorithmic designs of a free gyroscopic compass and FPS were suggested by measuring the earth's rotation rate on the basis of a free gyroscope (Jeong & Park, 2006;Jeong & Park, 2007).

This paper is to explain how to measure the nadir angle by using the earth's rotation rate. Firstly, the determination of the position on or near the earth is briefed. The motion rate of the spin axis caused by the earth's rotation rate is to be transformed into the platform frame and then into the local geodetic frame, i.e. the NED(north-east-down) navigation frame. Finally the nadir angle is to be obtained by using the rotation rate of the horizontal component on the NED navigation frame. And also, a free gyroscopic compass is explained by measuring the earth's rotation rate on the basis of a free gyroscope.

2 DETERMINATION OF VEHICLE'S POSITION BY NED NAVIGATION FRAME

First consider the transformation matrix C_i^n (Rogers R M, 2000) from the inertial frame to the navigation frame which is simply given by Eq. (1).

$$C_i^n = C_e^n C_i^e$$

$$= \begin{bmatrix} -\sin\phi\cos\lambda & -\sin\phi\sin\lambda & \cos\phi \\ -\sin\lambda & \cos\lambda & 0 \\ -\cos\phi\cos\lambda & -\cos\phi\sin\lambda & -\sin\phi \end{bmatrix} \begin{bmatrix} \cos\varpi_e t & \sin\varpi_e t & 0 \\ -\sin\varpi_e t & \cos\varpi_e t & 0 \\ 0 & 0 & 1 \end{bmatrix}$$

$$= \begin{bmatrix} -\sin\phi\cos(\lambda+\varpi_e t) & -\sin\phi\sin(\lambda+\varpi_e t) & \cos\phi \\ -\sin(\lambda+\varpi_e t) & \cos(\lambda+\varpi_e t) & 0 \\ -\cos\phi\cos(\lambda+\varpi_e t) & -\cos\phi\sin(\lambda+\varpi_e t) & -\sin\phi \end{bmatrix} \tag{1}$$

Here, ω_e is the (presumably uniform) rate of Earth rotation, λ is the geodetic longitude, ϕ is the geodetic latitude and t denotes time. This transformation matrix C_i^n denotes the transformation from the unit vectors of axes in the inertial frame to those in the navigation frame. Consider an arbitrary gyro vector $g_i^i = [\, u_x, u_y, u_z\,]^T$ which is unit vector in the inertial frame. We obtain easily the gyro vector transformed in the navigation frame, $g_u^n = [\, N_u, E_u, D_u\,]^T$ as Eq. (2).

$$g_v^n = C_i^n g_v^i \qquad (2)$$

$$= \begin{bmatrix} -\sin\phi\cos(\lambda+\varpi_e t) & -\sin\phi\sin(\lambda+\varpi_e t) & \cos\phi \\ -\sin(\lambda+\varpi_e t) & \cos(\lambda+\varpi_e t) & 0 \\ -\cos\phi\cos(\lambda+\varpi_e t) & -\cos\phi\sin(\lambda+\varpi_e t) & -\sin\phi \end{bmatrix} \begin{bmatrix} u_x \\ u_y \\ u_z \end{bmatrix}$$

$$= \begin{bmatrix} -u_x\sin\phi\cos(\lambda+\varpi_e t) - u_y\sin\phi\sin(\lambda+\varpi_e t) + u_z\cos\phi \\ -u_x\sin(\lambda+\varpi_e t) + u_y\cos(\lambda+\varpi_e t) \\ -u_x\cos\phi\cos(\lambda+\varpi_e t) - u_y\cos\phi\sin(\lambda+\varpi_e t) - u_z\sin\phi \end{bmatrix} = \begin{bmatrix} N_u \\ E_u \\ D_u \end{bmatrix}$$

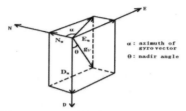

Fig.1 Measurement quantities in the navigation frame

As shown in Fig. 1, the azimuth angle of a gyro vector, α, and the nadir angle, θ, can be obtained respectively as Eq. (3) and Eq. (4), noting that $\because = \therefore$.

$$\cos\theta = \frac{U_x}{|g_v|} = -u_x\cos\phi\cos(\lambda+\varpi_e t) - u_y\cos\phi\sin(\lambda+\varpi_e t) - u_z\sin\phi \qquad (3)$$

$$\tan\alpha = \frac{E_u}{N_u} = \frac{-u_x\sin(\lambda+\varpi_e t) + u_y\cos(\lambda+\varpi_e t)}{-u_x\sin\phi\cos(\lambda+\varpi_e t) - u_y\sin\phi\sin(\lambda+\varpi_e t) + u_z\cos\phi} \qquad (4)$$

If we use two free gyros whose gyro vectors in Eq. (3) are $g_{va}^i = [u_{ax}, u_{ay}, u_{az}]^T$ and $g_{vb}^i = [u_{bx}, u_{by}, u_{bz}]^T$ respectively, we can determine the position (φ, λ) of a vehicle at the given nadir angles θa, θb. Once determining the position, we can also obtain the azimuth of a gyro vector by using Eq. (4). Park and Jeong(2004) already suggested the algorithm of how to determine a position.

3 SHIP'S HEADING, AZIMUTH AND NADIR ANGLE OF GYRO VECTOR

3.1 *Relation between ship's heading and azimuth of gyro vector*

As Jeong & Park(2006) mentioned, let's consider the earth's rate ϖ_e. Then its north component is $\varpi_e\cos\phi$, where ϕ depicts the geodetic latitude of an arbitary position. Fig. 2 shows that the angular velocities of the fore-aft and the athwartship components are given by Eq.(5) (Titterson, et al., 2004), where ψ is ship's heading. And it also shows that ς is the azimuth of a gyro vector from ship's head.

$$\omega_x = \varpi_e\cos\phi\cos\psi$$
$$\omega_y = -\varpi_e\cos\phi\sin\psi \qquad (5)$$

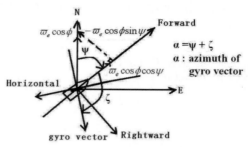

Fig. 2 Relation between ship's heading and azimuth of a gyro vector

By taking the ratio of the two independent gyroscopic measurement, the heading, ψ, is computed by Eq. (6).

$$\psi = \arctan\frac{\omega_y}{\omega_x} \qquad (6)$$

Meanwhile assuming that a gyro vector is ς away from ship's head, its azimuth from North is represented by Eq. (7). Therefore the angular velocity of the horizontal axis of a gyro (hereinafter called 'ϖ_H') is given by Eq. (8) on the navigation frame or local geodetic frame.

$$\alpha = \psi + \varsigma \qquad (7)$$

$$\omega_H = -\varpi_e\cos\phi\sin\alpha \qquad (8)$$

Eq.(8) shows that if the North component of the earth's rotation rate can be known on the navigation frame, the nadir angle of a gyro vector, θ, is obtained by Eq.(9), by integrating Eq. (8) incrementally over a time interval.

$$\theta = \int_{t_1}^{t_2}\varpi_H dt \qquad (9)$$

3.2 *Representation of the motion rate of the spin axis in the frames*

3.2.1 *The motion rate of the spin axis*

Let the motion rate of the spin axis in the gyro frame, $\omega_{i/g}^g = [0 \quad \omega_{gy} \quad \omega_{gz}]^T$, where we denote: $\varpi_{i/g} =$ the motion rate of the gyro frame(g) relative to the inertial frame(i), with coordinates in the gyro frame(g), and hereafter the same notation of the angular velocity is applied. In fact this angular velocity is all you can get from a free gyro and has to be transformed into the local geodetic frame through the platform frame or the body frame (Jeong & Park, 2006).

First, the motion rate of the spin axis in the gyro frame, $\omega_{i/g}^g$, is transformed into that in the platform frame, $\omega_{i/g}^p$, as follows.

$$\omega_{i/g}^p = C_g^p \omega_{i/g}^g \tag{10}$$

Next, the motion rate of the spin axis in the platform frame, $\omega_{i/g}^p$, is also transformed into that in the navigation frame, $\omega_{i/g}^n$, as Eq. (11).

$$\omega_{i/g}^n = C_p^n \omega_{i/g}^p \tag{11}$$

By the way assuming that there is no error in the free gyro and rate sensors, the motion rate of the spin axis in the local geodetic frame, $\omega_{i/g}^n$ (hereinafter called 'ω_L'), is equal to the angular velocity on the navigation frame, $\omega_{i/n}^n$, which is composed of "earth" and "transport" rates and rewritten in Eq. (13)(Rogers,2003). Considering that the earth's rotation rate, $\omega_{i/e}^n$, is shown in Eq,(14), it is represented by Eq. (13). Therefore the north component of the earth's rotation rate is computed by using the measured horizontal components, i.e. the fore-aft and athwartship ones.

$$\omega_L = \omega_{i/g}^n = \omega_{i/n}^n = \begin{bmatrix} \omega_{Lx} \\ \omega_{Ly} \\ \omega_{Lz} \end{bmatrix} \tag{12}$$

$$\omega_{i/n}^n = \omega_{i/e}^n + \omega_{e/n}^n$$

$$\omega_N = \omega_{i/e}^n = \omega_{i/g}^n - \omega_{e/n}^n = \begin{bmatrix} \omega_{Nx} \\ \omega_{Ny} \\ \omega_{Nz} \end{bmatrix} \tag{13}$$

$$\omega_{i/e}^n = \begin{bmatrix} \omega_e \cos\lambda \\ 0 \\ -\omega_e \sin\lambda \end{bmatrix} \tag{14}$$

The transport rate, $\omega_{e/n}^n$, is shown in Eq. (15). In Eq. (15) $\dot{\lambda}$ denotes the time rate of change of the longitude while $\dot{\phi}$ is the time rate of change of the latitude. And v_E is the east velocity, v_N is the north velocity, R is the radius of the earth and h is the height above ground.

$$\omega_{e/n}^n = \begin{bmatrix} \dot{\lambda}\cos\phi \\ -\dot{\phi} \\ -\dot{\lambda}\sin\phi \end{bmatrix} = \begin{bmatrix} \dfrac{v_E}{R+h} \\ -\dfrac{v_N}{R+h} \\ -\dfrac{v_E}{R+h}\tan\phi \end{bmatrix} \tag{15}$$

Meanwhile the ship's heading, ψ, can be computed by using Eq. (13) and given by Eq. (16). Of course Eq. (15) has to be transformed by multiplying the transformation matrix, C_n^H, which changes from the NE frame to the fore-aft and rightward one.

$$\psi = \arctan\frac{\omega_{Ny}}{\omega_{Nx}} \tag{16}$$

The azimuth of the gyro vector from the ship's head, ς, can be obtained by integrating the vertical component of the motion rate of the spin axis, i.e. Eq. (12) and given by Eq. (17).

$$\varsigma = \int_{t_1}^{t_2} \omega_{Lz} dt \tag{17}$$

The northward angular velocity of the local geodetic frame, ω_{LH}, which is determined by the sum of the rates the earth's rotation and the ship's transport, is represented by Eq. (18).

$$\omega_{LH} = \sqrt{\omega_{Lx}^2 + \omega_{Ly}^2} \tag{18}$$

And the horizontal component of the motion rate of the free gyro, ω_H, can be obtained by Eq. (19). It is evident that the azimuth of the gyro vector, α, is given by the sum of the ship's heading ψ and the azimuth of the gyro vector from the ship's head ς.

$$\omega_H = -\omega_{LH}\sin\alpha$$
$$\alpha = \psi + \varsigma \tag{19}$$

Therefore the nadir angle of the gyro vector, θ, can be obtained by integrating Eq. (19).

$$\theta = \int_{t_1}^{t_2} \omega_H dt \tag{20}$$

3.2.2 Coordinate transformation from gyro frame to platform frame

In Fig. 3 the gyro frame refers to free gyro itself on the platform, whose axes are defined along the spin(x_g), horizontal(y_g), and downward(z_g) directions. The platform frame refers to the vehicle to be navigated, whose axes are defined along the forward(x_p), right(y_p), and through-the-floor(z_p) directions.

The angle ξ is a rotation angle about the downward axis z_p and is positive in the counterclockwise sense as viewed along the axis toward the origin, while the angle η is a rotation angle about the horizontal axis(y_p) and is positive in the same manner as above. Here the transformation matrix C_g^p from the gyro frame to platform frame is given by Eq.(21), using Euler angles and direction cosines.

$$C_g^p = \begin{bmatrix} \cos\xi\cos\eta & -\sin\xi & \cos\xi\sin\eta \\ \sin\xi\cos\eta & \cos\xi & \sin\xi\sin\eta \\ -\sin\eta & 0 & \cos\eta \end{bmatrix} \tag{21}$$

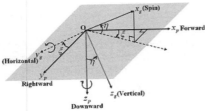

Fig. 3 Gyro and platform frames

3.2.3 Coordinate transformation into the NED navigation frame

With respect to the NED navigation frame whose axes are defined as the first axis points the north, the second axis points east and the third axis is aligned with the ellipsoidal normal at a point, in the downward direction. Let's consider the platform frame axes point forward, to the right, and down as shown in the above. Euler angles define the transformation, that is, they are the roll(R), pitch(P), and yaw(Y) relative to the NED axes as shown in Fig. 4. Then the transformation matrix C_p^n is given by Eq. (22).

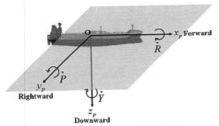

Fig. 4 Platform frame relative to NED frame

$$C_p^n = \begin{bmatrix} \cos P & \sin R \sin P & \cos R \sin P \\ 0 & \cos R & -\sin R \\ -\sin P & \sin R \cos P & \cos R \cos P \end{bmatrix} \quad (22)$$

3.2.4 Determination of transformation matrices

First, for transformation matrix C_p^g we have to know the rotation angles of the gyro frame, ξ and η. They are obtained by integrating the respective components of the spin motion rate. In doing so we can get the transformation matrix by solving the following first order linear differential equation (23) as Jeong & Park(2006) suggested.

$$\frac{dC_p^g}{dt} = -\Omega_{p/g}^g C_p^g \quad (23)$$

Here $\Omega_{p/g}^g$ is a skew-symmetric matrix and we assume it is constant over the sampling interval. The solution is given by Eq. (24).

$$C_p^g = \Psi(t,t_0)C_p^g(t_0)$$
$$\Psi(t,t_0) = \exp(\int_{t_0}^{t} (-\Omega_{p/g}^g)dt) \quad (24)$$
$$= I + \frac{\sin(|a|)}{|a|}A + \frac{1-\cos(|a|)}{|a|^2}A^2$$

$$A = \begin{bmatrix} 0 & a_3 & -a_2 \\ -a_3 & 0 & a_1 \\ a_2 & -a_1 & 0 \end{bmatrix} = \int_{t_0}^{t}(-\Omega_{p/g}^g)dt = -\Omega_{p/g}^g\Delta t$$

$$a_i = -\int_{t_0}^{t} \varpi_{p/g}^g(i)d\tau, \quad |a| = \sqrt{a_1^2 + a_2^2 + a_3^2}$$

Here $\Delta t = t - t_0$, t_0 is the initial time and a_i is each component of rotation angle. Once the transformation matrix C_p^g is obtained, the inverse matrix of it, C_g^p, is immediately calculated by doing the transpose of it since it is an orthogonal matrix. The relation between them is given by Eq. (25).

$$C_g^p = (C_p^g)^{-1} = (C_p^g)^T \quad (25)$$

Secondly, for transformation matrix C_p^n, we have to know the rotation angles of the platform frame P and R. They are obtained by integrating the respective components of the motion rate of the spin axis as shown in the above. We can also get the transformation matrix by solving the following first order linear differential equation (26).

$$\frac{dC_n^p}{dt} = -\Omega_{n/p}^p C_n^p \quad (26)$$

Here $\Omega_{n/p}^p$ is a skew-symmetric matrix too and we assume it is constant over the sampling interval. The solution is given by Eq. (27).

$$C_n^p = \Gamma(t,t_0)C_n^p(t_0)$$
$$\Gamma(t,t_0) = \exp(\int_{t_0}^{t} (-\Omega_{n/p}^p)dt) \quad (27)$$
$$= I + \frac{\sin(|b|)}{|b|}B + \frac{1-\cos(|b|)}{|b|^2}B^2$$

$$B = \begin{bmatrix} 0 & b_3 & -b_2 \\ -b_3 & 0 & b_1 \\ b_2 & -b_1 & 0 \end{bmatrix} = \int_{t_0}^{t}(-\Omega_{n/p}^p)dt = -\Omega_{n/p}^p\Delta t$$

$$b_i = -\int_{t_0}^{t} \varpi_{n/p}^p(i)d\tau, \quad |b| = \sqrt{b_1^2 + b_2^2 + b_3^2}$$

Here $\Delta t = t - t_0$, t_0 is the initial time and b_i is each component of rotation angle. Once the transformation matrix C_n^p is obtained, the inverse matrix of it, C_p^n, is immediately calculated by doing the transpose of it since it is an orthogonal matrix.

In addition, the other methods to solve the differential equations (23) and (26) are also represented by the integration of four quaternions or three rotation vectors, the integration of three Euler angle equations, and etc. Such equations suggested in the above are developed by referring to and using Farrell et al(1999), Jekeli(2001), and Rogers(2003).

4 ALGORITHMIC DESIGN OF FREE GYRO POSITIONING & DIRECTIONAL SYSTEM

Fig. 5 and Fig. 6 show the algorithmic design of free gyros positioning system mechanization. First, let's look into the ship's heading (Fig.5). In this mechanization two sensors for sensing the motion rate of the spin axis are mounted in the free gyro. Three sensors for sensing the motion rate of the platform are mounted in orthogonal triad. From the sensors in the gyro frame, the spin motion rate, $\omega_{i/g}^g$, is obtained and from the ones in the platform frame, $\omega_{i/P}^P$, is detected. By using the sum, $\omega_{P/g}^g$, of the rates from the free gyro and the ones detected from the platform sensors, the transformation matrix C_P^g is calculated and its inverse is determined. Therefore the spin motion rate, $\omega_{i/g}^g$, sensed from the free gyro is transformed into $\omega_{i/g}^P$ by using the inverse matrix, C_g^P.

Meanwhile the rate of the earth's rotation $\omega_{i/e}^n$ and the rate of the vehicle movement $\omega_{e/n}^n$ are summed and transformed into $\omega_{i/n}^n$. It is subtracted from the sensed rate from the platform, $\omega_{i/P}^P$. As a result, $\omega_{n/P}^P$ is generated. By using this, the transformation matrix, C_n^P, is calculated and the inverse of it, C_P^n, is obtained. And the rate $\omega_{i/g}^P$ is transformed into $\omega_{i/g}^n$ by using the transformation matrix, C_P^n. By using Eq. (13), the spin motion rate in the NED frame, ω_N, is obtained from the rate, $\omega_{i/g}^n$. Finally, <u>the ship's heading is calculated by using the components of the spin motion rate according to Eq. (13) and Eq. (16).</u>

Next let's look into the nadir angle (Fig.6). Because the motion rate of the spin axis in the local geodetic frame, $\omega_{i/g}^n$ is represented by ω_L, The azimuth of the gyro vector from the ship's head, ς, can be obtained by using Eq. (17). Then the azimuth of the gyro vector from the North, α, can be easily taken by Eq. (19).

The northward angular velocity of the local geodetic frame, ω_{LH}, is represented by Eq. (18). And the horizontal component of the motion rate of the free gyro, ω_H, can be obtained by Eq. (19). As a result <u>the nadir angle of the gyro vector, θ, can be obtained by Eq. (20).</u>

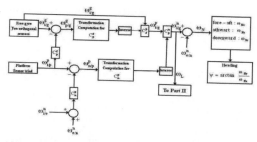

Fig. 5 Free gyro positioning system mechanization (1)

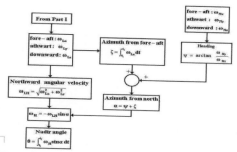

Fig. 6 Free gyro positioning system mechanization (2)

5 RESULTS AND DISCUSSIONS

This paper investigated and developed the algorithm regarding free gyro positioning system theoretically and analytically. As a result conclusions are the following.

1 Once the spin motion rate of free gyro is known, the ship's heading is determined by using Eq. (16).

2 The azimuth of the gyro vector from the ship's head, ς, can be obtained by Eq. (17). And the northward angular velocity of the local geodetic frame, ω_{LH}, can be given by Eq. (18).

3 The horizontal component of the motion rate of the free gyro, ω_H, can be obtained by Eq. (19). Finally the nadir angle of the gyro vector, θ, can be obtained by Eq. (20).

4 In order to transform the spin motion rate of the gyro frame into the one of the NED navigation frame, the differential equations of Eq. (23) and Eq. (26) are solved by using Eq. (24) and Eq. (27) and the transformation matrices are obtained respectively.

This paper ascertained the feasibility to set a stepping stone to the development of the free gyro positioning system. However, several problems remain unsolved in the aspect of the following. Firstly a two-degree-of-freedom gyro is very expensive and is commercially disadvantageous in practice. Secondly the inherent errors caused by many elements complicated. Errors caused by free gyro itself, sensors of the platform, sensors of the free gyro, sampling time and etc. will be dealt with in the next study.

REFERENCES

(1) Farrell, J. and Barth, M.(1999), "The Global Positioning System and Inertial Navigation", Mcgraw-hill, pp. 44-50.
(2) Jekeli, C.(2000), "Inertial Navigation Systems with Geodetic Applications", Walter de Gruyter, p.25.
(3) Jeong, T.G.(2005), "A Study on the Errors in the Free-Gyro Positioning System (I)", International Journal of Navigation and Port Research, Vol. 29, No. 7, pp. 611-614.
(4) Jeong, T.G. and Park, S.C.(2006), "A Theoretical Study on Free Gyroscopic Compass", International Journal of Navigation and Port Research, Vol. 30, No. 9, pp. 729-734.
(5) Jeong, T.G. & Park, S.C.(2007), "An Algorithmic Study on Free-Gyro Positioning System(I) - Measuring Nadir Angle by using the Motion Rate of a Spin Axis -", Journal of Navigation and Port Research, Vol. 31, No. 9, pp. 751~757.
(6) Park, S.C. and Jeong, T.G.(2004), "A Basic Study on Position Fixing by Free Gyros", Journal of Korean Navigation and Port Research, Vol.28, No.8, pp. 653-657.
(7) Rogers, R.M(2003), "Applied Mathematics in Integrated Navigation Systems", 2nd ed., AIAA, pp.65-66.
(8) Titterton, D.H. and Weston, J.L.(2004), "Strapdown Inertial Navigation Technology", AIAA, p. 287.

11. Compensation of Magnetic Compass Deviation at Single Any Course

E.M. Lushnikov
Szczecin Maritime Academy, Poland

ABSTRACT: The new method for compensation of deviation of magnetic compass at one any course is offered. The theoretical substantiation of a method is given, the analysis of accuracy is made, corresponding conclusions and recommendations are made. It allows to carry out a deviation's works without interruption from voyage.

1 INTRODUCTION.

The compensation of deviation of magnetic compass is usually carried out on the special aquatory equipped by leading line. The primary compensation of deviation is executed at an output of a vessel from building shipyard. All factors of deviation is determined and compensated in this case. The determination of residual deviation and calculation of the table is made after compensation of deviation. Such procedure can demand some hours of time. At annual deviation's works the compensation of the most inconstant factors of deviations B and C is made only. These factors on new building vessels can reach values $9^0 \div 12^0$. They are the most instable in storm conditions, at ice navigation, at knock about a quay on mooring, etc. As a rule, the table of deviation guarantees high reliability of the data up to the first heavy storm.

The most often used method for compensation of factors B and C is the method of Airy, which is carried out at 4 main magnetic courses. Accuracy of compensation depends on accuracy of supervision, on accuracy of operations by magnets - compensators, on hysteresis effects in the body of the vessel at maneuvering by means of course. After compensation of deviation the definition of residual deviation and calculation of the table is carried out.

Especially many problems are delivered at deviations maneuvers to large-capacity ships such as supertanker, big passenger ship, the big military ships and submarines etc.

Every time even the minimal program of deviation's work is connected to loss of operational time and an additional overhead charge. The problem of navigational safety is included in this case into the contradiction with economic problems. The radical decision of this question would be possible at presence of a method for destruction of deviation *without derivation of a vessel from the basic work*. Such statement of a question is possible only at presence of a method for destruction of deviation on one any course. The deviation's works at one course would allow as considerably to exclude influence of hysteresis effects on accuracy of deviation's works. Thus, the way of destruction of deviation on one any course is the most effective way to liquidate unproductive expenses of time.

2 THE DEVIATION OF MAGNETIC COMPASS AT CONTEMPORARY CONDITION.

At contemporary ships of symmetric design the constant factor of deviation A and the factor of deviation E depending from asymmetrical soft steel of the ship are in limits $0,2^0 \div 0,6^0$ and are characterized by extremely high stability [2]. The factor of deviation D after compensation by the help of without induction's sheet of a soft iron [1] does not exceed $0,25^0$ and as differs very high stability.

It can to tell, that the values of these three factors of deviation are situated at the same level as accuracy of supervision of courses and bearing. However, according to rigid algorithm of Airy, these factors without any need are determined and recalculated anew for use in the new table of deviation [3].

All this operations can be qualified, as unproductive works with loss of time for measurements, processing and calculations.

Exact expression for deviation of a magnetic compass δ is implicit function from compass course KK and enters the name as:

$$Sin\delta = A\cos\delta + BSinKK + CCosKK + DSin(2KK + \delta) + ECos(2KK + \delta)$$
(1)

where:

$$A = \frac{d-b}{2\lambda}; \quad B = \frac{P+cZ}{\lambda H}; \quad C = \frac{Q+fZ}{\lambda H}; \quad D = \frac{a+e}{2\lambda}; \quad E = \frac{d+b}{2\lambda}$$

thus:

H - a horizontal component of force of terrestrial magnetism;

Z - a vertical component of force of terrestrial magnetism;

P, Q - longitudinal and cross-section magnetic forces from according hard ship's steel;

a, b, c, d, e, f - parameters of Poisson, describing constructions from soft ship's steel;

$$\lambda = 1 + \frac{a+e}{2}$$ - factor of shielding of a magnetic compass.

Parameters of Poisson a, b, c, d, e, f and as factor λ, are functions of the sizes and forms of ship's soft steel, his remoteness from a compass and magnetic characteristics of a case material. All these characteristics are constant constructive parameters of a vessel, than high stability of factors A, D, E explains.

Taking into consideration this circumstance, factors of deviation A, D, E usually consider constant and at performance of annual procedural works these factors do not adjust. In this case the problem of annual deviation's works is reduced to indemnification of factors B and C and to calculation of the new table of deviation. Such operations at annual deviation's works are the established practice already for a long time.

The last ministry's instruction of Russia "Recommendations to navigation's service" of 1989 year do not define the time of actuality for a table of deviation. Only the level of accuracy according to requirements of IMO is formulated at this instruction. At the same time "Recommendations to navigation's service for a ships of a fishing fleet" contains record about the maximal 1 year interval of actuality of the deviation's table. These departmental distinctions emphasize complexity and a urgency of this problem.

Progress in development of satellite systems of navigation and gyrocompasses has led to that magnetic compasses on sea vessels basically carry out reserving and monitoring function. Unproductive expenses of time for deviation's works stimulates a negative attitude of ship-owners and captains of ships.

Modern market conditions demand optimization of production and the proved time expenses. It is natural, that such optimization should be made in view of safety of navigation.

3 PRECONDITIONS TO DESTRUCTION OF DEVIATION WITHOUT INTERRUPTION OF VOYAGE.

If the factors of deviation A, D, E are small and constant, there is no need to spend time for determination of these factors anew. It is necessary to take into account their values from the previous table.

The same logic can be continued further. Factors B and C at carrying out of deviation's work can be not destroyed up to zero, and to restore their former residual tabulated values [4].

Such step gives the basis to consider, that after restoration of factors B and C all factors of deviation correspond(meet) to values of the old table of deviation and to expect the new table there is no necessity.

Validity of the former table in this case can be prolonged for one year. All deviation's works will be reduced in this case only to restoration of factors B and C without expenses of time for 8 courses for determination and calculation of all five factors. Also there is not necessity for calculation of new deviation's table . Such actualization of the former table of deviation can be made during 4÷5 years.

However the determination of factors B and C for the purpose of their return to former tabulated values demands not less than two equations, that is, at least, two courses. Otherwise it means, that compensation of two factors B and C at one course is impossible.

It is possible to notice, however, that in navigating practice exists essentially various two ways of determination of deviation. The first way bases on use of navigating measurements. The second way bases on physical measurements of magnetic forces with the subsequent calculation on this basis of deviation's factors.

Simultaneous use of these two essentially various methods allows to receive the missing information for the determination of a task in view on destruction of two factors deviations B and C at one course.

4 DETERMINATION OF FACTORS B AND C AT ONE ANY COURSE.

The set *of navigating* ways and means for determination of deviation of a magnetic compass on an any course of a vessel is known. For this purpose it is possible to use a terrestrial leading line, celestial object, remote reference points, systems AIS and gyrocompasses. The deviation of a magnetic compass δ determined by navigating way can be written down as implicit function of compass course KK as expression 1.

Taking into account, that in terms of 1 set sizes are deviation δ (measured by navigating way), compass course KK, and as factors A, D and E (from the previous table), the expression 1 can be copied to more compact kind:

$$B \sin KK + C \cos KK = \Delta_1 \qquad (2)$$

where:

$$\Delta_1 = \sin\delta - A\cos\delta - D\sin(2KK+\delta) - E\cos(2KK+\delta) \quad (3)$$

Thus, the equation 2 connects two unknown factors of deviation B and C by means of measurement of deviation δ.

As the second missing equation can be used equation of total ship's magnetically force of compass H_K. It is known [2], that the value of measured force H_K looks like:

$$H_K = \lambda H[\cos\delta + A\sin\delta + B\cos KK - C\sin KK + D\cos(2KK+\delta) - E\sin(2KK+\delta)] \quad (4)$$

Expression 4 can be copied to more compact kind:

$$B \cos KK - C \sin KK = \Delta_2 \qquad (5)$$

where

$$\Delta_2 = \frac{H_k}{\lambda H} - \cos\delta - A\sin\delta - D\cos(2KK+\delta) + E\sin(2KK+\delta) \qquad (6)$$

Thus, the system of two equations 2 and 5 at two unknown factors B and C is received:

$$\begin{aligned} B \sin KK + C \cos KK &= \Delta_1 \\ B \cos KK - C \sin KK &= \Delta_2 \end{aligned} \qquad (7)$$

The solution of this system of the equations gives:

$$\begin{aligned} B &= \Delta_1 \sin KK + \Delta_2 \cos KK \\ C &= \Delta_1 \cos KK - \Delta_2 \sin KK \end{aligned} \qquad (8)$$

At essential changes of these factors they must be restoring by means of regulators B and C of compass before former table's values. For restoration of former values of factors B and C the value of correction ΔB and ΔC is calculated under formulas:

$$\begin{aligned} \Delta B &= B_T - B \\ \Delta C &= C_T - C \end{aligned} \qquad (9)$$

where B_T and C_T - values of factors B and C from the table of deviation.

If factors of correction ΔB and ΔC are positive, readout of each regulator increases before the corresponding value and on the contrary.

Thus, joint application of navigating and physical measurements allows to solve a problem which all time was considered insoluble.

Both factors B and C depend from correction a component Δ_1 and Δ_2. Navigating component Δ_1, ap-

parently from expression 4, depends on accuracy of definition of deviation δ and from accuracy of tabulated factors A, D, E. Correction component Δ_2, apparently from expression 7, demands knowledge of exact values of resulting compass force H_k, a horizontal component of terrestrial magnetism H, factor λ, and as deviation δ and factors A, D, E. Except for accuracy of the navigating data the exact data of physical measurements here are required. Accuracy of attitude H_k/H can be provided with use of the same deflector for measurements on coast and on a vessel.

Accuracy of factor λ in usual circumstances never represented special interest. In this case of accuracy of knowledge of this factor are demanded much.

The situation is facilitated by that it needs to be determined accuracy once as his stability as is extremely high as stability of factors A, D, E.

Believing, that deviations are characterized by rather small angles, that usually corresponds to the validity, both settlement components Δ_1 and Δ_2 at high accuracy can be simplified to a kind:

$$\begin{aligned} \Delta_1 &= \delta - A - D\sin 2KK - E\cos 2KK \\ \Delta_2 &= \frac{H_k}{\lambda H} - 1 - D\cos 2KK + E\sin 2KK \end{aligned} \qquad (10)$$

In view of these simplifications the correction ΔB and ΔC will become:

$$\begin{aligned} \Delta B &= (\delta - A + E)\sin KK + \left[\frac{H_k}{\lambda H} - 1 - D\right]\cos KK \\ \Delta C &= (\delta - A - E)\cos KK - \left[\frac{H_k}{\lambda H} - 1 + D\right]\sin KK \end{aligned} \qquad (11)$$

Final record of factor ΔB and ΔC can be submitted as:

$$\begin{aligned} \Delta B &= (\delta - M)\sin KK + \left[\frac{H_k}{\lambda H} - N\right]\cos KK \\ \Delta C &= (\delta - U)\cos KK - \left[\frac{H_k}{\lambda H} - V\right]\sin KK \end{aligned} \qquad (12)$$

where:

$$M = A - E;$$
$$N = 1 + D;$$
$$U = A + E;$$
$$V = 1 - D.$$

Factors M, N, U, V it is necessary to calculate at once after full indemnification of deviation and calculation of the table of residual deviation. Formulas 12 and value of factors M, N, U, V are used at the further annual procedural works on compensation of deviations factors B and C.

Substitution of these numerical values in beforehand prepared formulas allows to calculate quickly values of correction's factors ΔB and ΔC and to enter them with the help of corresponding regulators.

Application of such method directly at a cargo mooring, as a rule, is not expedient owing to presence on a mooring and in designs of a mooring of the big iron weights, and as positions of ship iron not in a marching way.

The method is the most expedient for applying at an output of a vessel from port when it is situated on leading line. Such operation can be executed by deviator so as ship's navigator. For performance of works it is required no more than 10 minutes. In this case disappears necessity of special aquatory and additional time for deviation's work.

All this process can be named as a process of restoration or process of actualization of the former table of deviation. The most important in all it is that this actualization can be made on one any course without interruption of voyage.

5 THE ANALYSIS OF ACCURACY OF A METHOD

It is obvious, that accuracy of restoration of the table of deviation depends on accuracy of determination of proof values ΔB and ΔC. They, in turn, depend on accuracy of measurement of deviation δ, from accuracy of the information about tensions of magnetic fields H_K and H, and as from accuracy of factor λ.

Regular error of actualization of deviation's table. For an estimation of a regular error of restoration of the table of deviation it is necessary to execute differentiation of expressions (11) therefore it turns out:

$$d\Delta B = d\delta \cdot \sin KK + \left[\frac{\lambda H dH_k - HH_k d\lambda - \lambda H_k dH}{\lambda^2 H^2} \right] \cdot \cos KK$$
$$d\Delta C = d\delta \cdot \cos KK - \left[\frac{\lambda H dH_k - HH_k d\lambda - \lambda H_k dH}{\lambda^2 H^2} \right] \cdot \sin KK \qquad (13)$$

Believing, that measurement of force H on coast and force H_K on a vessel was made by means of the same deflector and by the same observatory these measurements can be qualify as the same accuracy.

$$dH = dH_k$$

In this case expression (13) corresponds to a kind:

$$d\Delta B = d\delta \sin KK + \left[\frac{(H-H_k)dH}{\lambda H^2} - \frac{H_k d\lambda}{H\lambda^2} \right] \cdot \cos KK$$
$$d\Delta C = d\delta \cdot \cos KK - \left[\frac{(H-H_k)dH}{\lambda H^2} - \frac{H_k d\lambda}{H\lambda^2} \right] \cdot \sin KK \qquad (14)$$

Apparently from expression (14), accuracy of restoration of the table of deviation depends on accuracy of a navigating component of measurements $d\delta$, a technical component of measurements dH, and also an information component $d\lambda$.

For estimating calculations it is possible to count that $H \approx H_K$, $\lambda \approx 1$. In view of told, for an estimation of

accuracy as a first approximation expression (14) can be simplified to a kind:

$$d\Delta B = d\delta \cdot \sin KK - d\lambda \cos KK$$
$$d\Delta C = d\delta \cdot \cos KK + d\lambda \cdot \sin KK \qquad (15)$$

From this expression it is visible, that the main factors of regular errors are accuracy of navigating supervision and accuracy of knowledge of factor λ. The regular error of determination of deviation at leading line is extremely small. In this connection the basic role belongs to a component depending on factor λ. For maintenance of accuracy at a level 0.5^0 relative error of factor λ should not exceed 0,8 %. Such requirement is high enough, but quite real. Determination of factor λ is carried out by measurement of compass force H_k on four main and four intermediate course's with the subsequent calculation under the formula:

$$\lambda = \frac{\sum_1^8 H_k}{8H}$$

The requirements of Register to accuracy of compensation of deviation is $\delta \leq 3^0$. The relative *methodical* error of determination of factor λ will be not worse, than 0,12 % . Such accuracy is more than sufficient.

Exact value of factor λ should be determined at descent of a vessel to water. The information on factors A, D, E and as about factor λ it should be kept carefully on a vessel before the next complex check and compensation of deviation. At capital reconstruction of a vessel, replacement of the engine these factors should be determined anew.

Casual errors of actualization of the table of deviation. Influence of casual errors of supervision and measurements is estimated by the help of standard error under the formula:

$$m_x = \sqrt{ \left(\frac{\partial f}{\partial x_1} \right)^2 \cdot m_{x_1}^2 + \left(\frac{\partial f}{\partial x_2} \right)^2 \cdot m_{x_2}^2 + ... + \left(\frac{\partial f}{\partial x_n} \right)^2 \cdot m_{x_n}^2 }$$

Using as function f expressions (11), we shall receive standard errors of the proof data ΔB and ΔC as:

$$m_{\Delta B} = \sqrt{ m_\delta^2 \sin^2 KK + \left[\frac{m_{Hk}^2}{\lambda^2 H^2} + \frac{H_k^2 m_H^2}{\lambda^2 H^4} + \frac{H_k^2 m_\lambda^2}{\lambda^4 H^2} \right] \cos^2 KK }$$

$$m_{\Delta C} = \sqrt{ m_\delta^2 \cos^2 KK + \left[\frac{m_{Hk}^2}{\lambda^2 H^2} + \frac{H_k^2 m_H^2}{\lambda^2 H^4} + \frac{H_k^2 m_\lambda^2}{\lambda^4 H^2} \right] \sin^2 KK } \qquad (16)$$

For estimated calculations it is possible to accept $H_k \approx H; \lambda \approx 1;$. At such assumptions of expression (16) become simpler to a kind:

$$m_{\Delta B} = \sqrt{m_\delta^2 \sin^2 KK + \left[\left(\frac{m_{H_k}}{H_k}\right)^2 + \left(\frac{m_H}{H}\right)^2 + (m_\lambda)^2\right]\cos^2 KK}$$

$$m_{\Delta C} = \sqrt{m_\delta^2 \cos^2 KK + \left[\left(\frac{m_{H_k}}{H_k}\right)^2 + \left(\frac{m_H}{H}\right)^2 + (m_\lambda)^2\right]\sin^2 KK} \tag{17}$$

From these expressions it is visible, that casual errors of compensation of factors B and C depend on relative errors of all three factors – navigating, technical and information.

At standard error of deviation at the level $m_\delta = 0,5^0$, at relative accuracy of magnetic forces at the level of 1 % and at relative accuracy of factor λ also at the level of 1 % a standard errors ΔB and ΔC is not lower 1^0. Schedule of standard errors $m_{\Delta B}$ and $m_{\Delta C}$ for such initial data is submitted in figure 1.

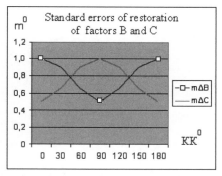

Fig. 1 The standard errors and depending from compass course at $m_{\Delta B}$ $m_{\Delta C}$ $m_\delta = 0,5^0$ and

$$\frac{m_H}{H} = \frac{m_{H_k}}{H_k} = \frac{m_\lambda}{\lambda} = 0,01$$

From figure it is visible, that casual errors of restoration of factors B and C are in limits $0,5^0 \div 1,0^0$.

The additional errors from instability of factors A, D and E are small, and stability of them is very high. Such accuracy of actualization of deviation's table is quite sufficient.

Not always the innovation gives a prize without by-effects and additional expenses. This case just does not entail any additional questions and problems.

6 THE CONCLUSION

1 The offered method for compensation of deviation of a magnetic compass on one any course of a vessel is essentially new method allowing to reduce a routine work of a vessel, connected with financial expenses.
2 The method differs exclusive simplicity. It can be applied by navigators in conditions of voyage.
3   For introduction of a method in practice of navigation it should find reflection in corresponding program of educational institutions.

THE LITERATURE

1. V.V. Voronov, N.N. Grigoriev., A.V. Jalovenko. Magnetically compass. Sankt-Petersburg. "ALMOR", 2004.
2. Kozuchov V.P., Voronov V.V, Grigoriev N.N. Magnetically compass. Moskov.: Transport, 1981.
3. E.M. Lushnikov. Compensation of magnetic compass deviation at contemporary conditions. International scientific conference «Innovation in scientific and education –2008» Kaliningrad, KGTU, 2008.
4. E.M. Lushnikov. The problem of magnetic compass deviation at contemporary condition. International Navigational Symposium "TRANSNAV 09". Gdynia, Maritime. University 2009. p.219-224.

Navigational Simulators

12. New level of Integrated Simulation Interfacing Ship Handling Simulator with Safety & Security Trainer (SST)

K. Benedict, C. Felsenstein & O. Puls
Hochschule Wismar, Dept. of Maritime Studies, Rostock-Warnemuende, Germany

M. Baldauf
World Maritime University, Malmoe, Sweden

ABSTRACT: Simulators have proved beneficial for ship handling training in real time on well equipped bridges throughout the last decades. The Maritime Simulation Centre Warnemuende (MSCW) has been complemented by a new type of simulator called the Safety and Security Trainer (SST). Wismar University has been involved in the conceptual design and development of this new technology. One of the most challenging innovations developed during the research is the 3D-designed RoPax ferry "Mecklenburg-Vorpommern" for the SST simulation system. An integrated support and decision system, called MADRAS, was interfaced into the SST and the entire system was interfaced to the Ship Handling simulator SHS in order to assists officers in coping with safety and security challenges during manoeuvres of the vessel (SHS). This new and enhanced simulation facility allows for "in deep" study of the effects of the safety and security plans and procedures on board and enable more detailed evaluation of their effectiveness under varying conditions and during different courses of events by a different series of simulation runs. This paper will introduce the basic concept of the safety and security training simulator and describe the work entailed for its integration into the complex environment of full mission ship-handling-simulators. Selected results of a case study dealing with first basic implementation of training scenarios will be demonstrated.

1 INTRODUCTION - INTERNATIONAL REGULATIONS FOR MARITIME SAFETY & SECURITY - LEVELS OF COMPETENCIES AND SIMULATOR TRAINING

The Diplomatic Conference on Maritime Security in London in December 2002 adopted new provisions in the International Convention for the Safety of Life at Sea, 1974 and the International Code for the Security of Ships and of Port Facilities - ISPS Code, which came into force 01st July 2004. The Code is in two parts, Part A which is mandatory and Part B which is recommended. The minimum requirements for ships respectively ports are ship (port facility) security assessment, ship (port facility) security plans in ports and on board the vessels and certain security equipment. Apart from existing regulations it is very important to recognize the importance of permanent process in changing and developing precautions and measures implemented to fight terrorism in port and on board the vessel. Human mental attitudes and motivation are important and necessary for to creating a general atmosphere of security culture.

The situation in the shipping world with regard to emergency preparedness is affected in general by the following elements:

– Abilities and Experiences in case of „disturbed" operation of systems are reduced or simply not existing
– Multilingual Crews cause specific problems in case of Emergency Situation
– Reduction of Crew Members causes lack of available Personnel
– Complexity of Emergency Equipment is permanently increasing, but Training in Emergency Handling has not developed to the same standard
– New Management Systems and regulations of the IMO (ISM/ISPS) demand new methods and technology for emergency training

According to the demand for increased level of training (see Figure 1) along with the requirements for higher competency level the simulator equipment at Dept. of Maritime Studies of Hochschule Wismar was extended: Additionally to the existing simulators at the Maritime Simulation Centre Warnemuende a new Safety and Security Trainer was implemented and interfaced to allow for the training on the highest level for the management level for integrated training with full mission simulators in interfaced mode of operation.

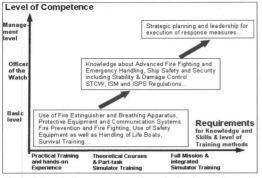

Figure 1. Level of competence and required safety and security training

2 INTEGRATED SIMULATION AT THE MARITIME SIMULATION CENTRE WARNEMUENDE (MSCW) WITH NEW ELEMENT SST

The Maritime Simulation Centre Warnemuende (MSCW) is one of the most modern simulation centres worldwide. The complex simulation platform (Figure 2; Benedict 2000) with several full mission simulators enables the department to simulate the entire "system ship" with the maritime environment including VTS and offers challenges to officers and crew on board the vessels (http://www.sf.hs-wismar.de/mscw/). The simulator arrangement (MSCW) comprises already

– a Ship Handling Simulator SHS with for 4 Full Mission bridges and 8 Part Task Bridges,
– a Ship Engine Simulator SES with 12 Part Task station and
– a Vessel Traffic Services Simulator VTSS with 9 operator consoles

The new simulator, implemented as Safety and Security Trainer SST, was designed by the manufacturer Rheinmetall Defence Electronics Bremen in co-operation with Wismar University, Department of Maritime Studies (Benedict et al 2008, Oesterle 2007). The simulator was originally designed in a basic version and 2D presentation and is now being developed into a 3D version. The simulator can specifically be used for stand alone and for integrated training with the SHS (Figure 3). Beside the use for training, the simulation system will be installed and used also for specific simulation based studies into potential upgrading of existing safety and security procedures.

3 WORKPLACE CONCEPT OF SAFETY- AND SECURITY TRAINER (SST)

10 stations are being installed in the MSCW this year, eight training stations (one of the stations on the SHS Bridge 1) and two instructor consoles as well as one communication computer system and another computer for a new support and decision system called MADRAS. Each station (with head phones or microphone for communication) consists of two monitors. One screen is used as Situation Monitor and the other is named Action Monitor. The workplace concept provides full equipment for comprehensive safety and security training (Figure 2, right).

A person simulating a member of the crew can be moved by mouse clicks through the decks on the situation monitor. The name of selected person, health index and moving type (standing, kneeing and lying) is shown in the status display window, also the kind of protective clothes worn by the figure.

Positioning the figure close to a consol the related safety equipment is indicated as generic panel on the action Monitor. All interaction is done on the action monitor. If the acting person is not located close to consoles or instruments representing safety equipment, the action monitor shows the ship safety plan of the appropriate deck.

For the instructor it is possible to create new or editing existing exercises and store replays. Also malfunctions, fire, water inrush and criteria for the incorporated assessment can be set.

Integration of Fire Fighting System and Fire Fighting Equipment: Most of the actions performed by the trainees with the safety equipment are performed on the action monitor. A fire model optimised visually and given obvious realistic effects for easy perception by trainees, is incorporated into the simulator. A modern fire alarm management system with smoke detectors and manual calling points is built into the interior of the ship and easily flammable materials are protected by fire resistant A60 walls and doors.

The fire model includes smoke visualisation and a fire fighting system and equipment such as fire extinguishers, water hoses and hydrants, breathing apparatus, CO_2 systems and foam. This enables the trainee to simulate a realistic fire fighting situation on board and interact with supporting teams as well as the management team on the bridge and in the engine room.

Figure 2. Overview on MSCW (left), Bridge 1 of Ship-Handling-Simulator (SHS) with new Displays of Bridge Safety & Security Centre of SST and MADRAS Decision Support System (right top) and Training room of new Safety & Security Trainer of SST (right bottom)

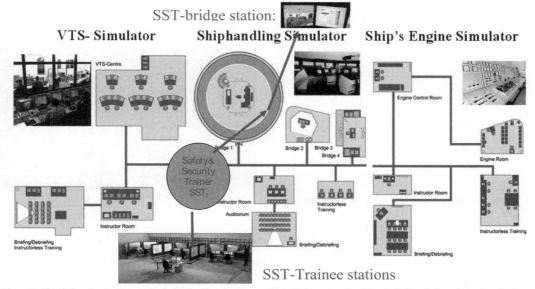

Figure 3. Simulation Centre Warnemünde (MSCW) – structure and interfacing network with new Safety & Security trainer SST

During the simulation the persons' health condition is monitored in relation to oxygen, smoke, temperature and other health influencing parameters and the measurements are monitored in diagrams

Integration of Water Inrush System: One feature of the simulation system is a model calculating water inrush and its influence to the stability of the ship. A ballast system is implemented and can be used during simulation of an emergency instance to help stabilize the ship. The trim and stability calculator is used to predict the effect of a water inrush and show the stability, bending moments and share forces. Water tight doors are built into the modelled vessel. The ballast and stability measuring system is implemented in the simulator, which enables the trainee to take countermeasures.

4 SPECIFIC SIMULATION FEATURES FOR THE RESEARCH PROJECT "VESPER"

4.1 Elements of the Research Project "VESPER"

The research project "VeSPer" is dedicated to the "Enhancement of passengers' safety on RoRo-Pax-ferries" and was designed thanks to various initiatives from the German government such as "Research for civil safety" and specifically "Protection of traffic infrastructures". The project is supported by the Ministry of Education and Research, under the aegis of the Technology Centre Düsseldorf (VDI). One of the most challenging innovations developed during the research is the implementation of the 3D-designed RoPax ferry "Mecklenburg-Vorpommern" for the SST simulation system.

The focus of investigations within the project "VeSPer" is laid on
- check-in procedures to increase the safety level for entrances to ferry ships and ports
- preventive measures on board (constructive and administrative)
- Sea side protection of ships in ports as well as in open sea when sailing
- investigations into potential improvement of measures in the case of a crisis
- The analysis and investigations deal with subjects such as:
- use and optimisation of monitoring and detection systems
- aspects of potential integration of decision support systems on board ships
- identification of potential for optimisation of processes and measures/procedures including the integration of new innovative technologies and
- consideration and application of rules and regulations according to national and international law

With reference to risk based scenarios in ports and on board the vessels following investigations are processed

- Process Analysis from entering the port, including booking and check in procedures, on approaching access to the vessel and access of embarkation
- Process Analysis on board the vessel from embarkation/departure until arrival/disembarkation
- Analysis of the ISPS Code and measures for the full integrated application on board
- Measurements for improved processes on board and access to the vessels and developing new security technologies and procedures
- Development of a support decision system for emergency measures on board the vessel in case of safety and/or security casualties

4.2 Integration of innovative 3D-visual model of SST

One of the most interesting innovations at the MSCW – apart from recent investments to technically upgrade the system of the SHS which marks a further noteworthy improvement and underlines the position of the MSCW as the leading simulation institute in Europe – is the 3D-designed RoPax ferry "Mecklenburg-Vorpommern" for the SST.

The first step was to make an application of the ship plans which were intricately realised in a 3D Studio Max version by HSW for test trials of the spectacular 3D-visualisation of the entire vessel. All decks of the RoPax ferry are now available in the 3D-version and integrated along with the dynamic safety equipment into the games engine by RDE. Functional tests of the developed system are in progress and already running successfully. Figure 4 and Figure 5 show the 3D visualisation of decks and public areas of the ferry.

Figure 4. Deck 9 of the RoPax ferry in 3D visualisation

Figure 5. Public area of the RoPax ferry in 3D visualisation

4.3 Safety and Security Components in the 3D Visualisation Model

In the 3D model moves and reacts from his own perspective and can operate the entire spectrum of safety equipment on board the vessel. In the case of fire he activates the alarm from the next manual calling point. According to the safety procedure on board, and after the release of the fire alarm from the bridge, the fire squad team (each trainee with specific role) will operate the fire fighting equipment including the breathing apparatus, fire protection suits, fire extinguishers, fire hoses and other tools located in the safety lockers or placed in the fire boxes (Figure 6 and Figure 7).

Figure 6. Fire fighting / smoke propagation in public area on deck 5 RoPax ferry

Figure 7. Crew in action with fire fighting equipment car deck 5 RoPax ferry

On the bridge (Figure 8) and in the engine control room (ECR - Figure 9) all the operational consoles including; steering panel, fire panel, alarm panel, ballast- and stability panel and the water drenching system, are designed to a generic model and can be integrated on other designed vessels as well. All consoles and panels on the bridge and in the ECR correspond to the integrated sensors placed all over the vessel. The Master and officers operate an interactive board system and can be trained in a wide spectrum focussing on safety and security procedures.

Figure 8. Bridge and interactive consoles

Figure 9. Engine control room with interactive consoles

In addition, the security components can be practised on the new simulator. For example the RFID based appliance, which is integrated into the SST bridge station, enables the officer to observe the movement of persons on board. In all security declared areas the doors are locked and the areas are accessible only by entering the specific code into the lock system beside the doors. On all decks cameras are installed and can be monitored from the bridge station. The camera view can be changed and adjusted by the instructor.

In the case of a bomb alert the crew can investigate the affected area with a bomb detector. On approaching any dangerous object, the detector sounds alarm. Figure 10 shows a crewmember crawling in the direction of a suspicious suitcase. When the bomb has been identified the dangerous object can be removed with a new remote controlled defence system called TELEMAX. This multipurpose vehicle can be used to detect and approach any suspicious objects from a safe distance using the remote control.

The threat of gas attack has also been integrated into the simulation system of mars³. In this kind of a threat the crew could approach the affected area wearing protection suits and breathing apparatus and can undertake all appropriate measures, i.e. for ventilation and evacuation of passengers.

Figure 1 Bomb search in the lounge and removal of suspicious object

4.4 Support and Decision System MADRAS

The simulation platform includes a new support and decision system called MADRAS. This system was designed by the company MARSIG mbH Rostock and especially tailored for the SST simulator and the simulated RoPax Ferry "Mecklenburg-Vorpommern". The MADRAS computer is linked to the mars³ simulator and receives the sensor data from the SST. The control module selection contains the following elements for automatic survey; FIRE, EXPLOSIVES, SECURITY, EVACUATION, GROUNDING and FLOODING. In the event of any sensor alarm the Madras menu opens and displays the affected deck/area with the activated alarm sensor. The following menus can be selected:

MONITORING – list of all existing sensors, grouped in different types and presenting the actual data of sensors

DECISION SUPPORT – recommendation structure and decision advise in specific safety- and security issues including necessary procedures:

OVERVIEW– deck overview displaying all installed sensors and highlighting the activated ones including diagrams

DEVICE CONTROL – list of all sensors – according to type, location, showing maximum and minimum values and the adjustable alarm level

PROTOCOL CHECK– date and time of sensor activation, location loop of sensors, duration of alarm, values of alarm and time record for reset

CONTROL – menu for sensor connections, support manager, value input, extended functions and system options

MADRAS is an interactive system and is a helpful tool for Master and officers in critical situations. The system guides the officer through all necessary choices and helps in finding the correct emergency procedures. This helps to avoid dangerous mistakes and ensures not missing any steps imperative for the safety of the vessel.

MADRAS was recently installed into the SST and is still under development. Test trials are running successfully. The basic system of MADRAS was tested on board of the ferry "Mecklenburg-Vorpommern" during the last two years.

5 SUMMARY AND CONCLUSIONS

Within the frame of investigations into potential enhancements of maritime safety and security the use of simulation facilities were investigated. The Safety and Security Trainer SST is a new product developed by Rheinmetall Defence Electronics (RDE) Bremen in co-operation with the Wismar University, Department of Maritime Studies in Rostock-Warnemuende. It can be operated in a standalone version for up to eight training stations and could be extended to include the training of the entire crew. The SST is also designed for integration into complex systems and was interfaced now with the existing ship handling simulator SHS of the MSCW for training of comprehensive scenarios in combination with the SHS, SES und VTS. The complex simulation platform with the full mission simulators enables the trainees to simulate the entire ship system and presents challenges to both officers and crew. A new quality of scenarios can be generated now for the comprehensive training of ship officers. On the other hand this new and enhanced simulation facility allows for in depth studies of the effects of ship's safety procedures and to evaluate their efficiency.

ACKNOWLEDGEMENTS

The investigations and developments described here are mainly performed in a project for research and technical development funded by the German Ministry of Education and Research Berlin and surveyed by VDI Technology Centre Düsseldorf. During the project also cooperation were established with World Maritime University. This cooperation covers e.g. aspects of international harmonisation of training requirements and standards. The authors would like to thank Rheinmetall Defence GmbH as well as the company AIDA Cruises Ltd and the involved ferry companies TT-Line and Scandlines for their grateful assistance and cooperation.

REFERENCES

Benedict, K. (2000) Integrated Operation of Bridge-, Engine Room- and VTS-Simulators in the Maritime Simulation Centre Warnemuende. Conference on Simulation - CAORF 2000, New York, 3-7 July 2000, Proceedings Vol. 1.

Benedict K., Felsenstein C., Tuschling G., Baldauf M. (2008) New Approach for Safety and Security Training in mars[2] Simulator. in: Proceedings of 35th International Marine Simulator Forum (IMSF), Warnemünde, 08.-12. Sept. 2008 – ISBN 978-3-939159-55-1

Oesterle, A. (2007). New Simulator Safety and Security Trainer mars[2] and its use for training. (in German) in: Moderne Konzepte in Schiffsführung und Schifffahrt. Schriftenreihe des Schifffahrtsinstituts, Rostock, 2007, Vol. 7.

13. Path Following Problem for a DP Ship Simulation Model

P. Zalewski
Maritime University of Szczecin, Szczecin, Poland

ABSTRACT: For the dynamic positioning (DP) operations the equations of ship's motion can be simplified to a 3 degrees-of-freedom (DOF) model. DP station-keeping is a set-point regulation problem, i.e., forcing the output of the ship to maintain a constant reference position and heading. However since offshore vessels reposition themselves at different locations, the path following problem has to be solved. The article presents various solutions of this problem including fuzzy control design which could be utilised in autonomous ship simulation as well.

1 INTRODUCTION

The Auto Track DP modes enable the vessel to follow a predefined path, described by a set of waypoints, with a high degree of accuracy [Kongsberg, 2008]. These modes cover both low-speed and high-speed operations using different control strategies. The system can automatically switch between the strategies depending on the requested speed.

Figure 1. DP MSV (multipurpose supply vessel) in Auto Track mode. [Kongsberg, 2008]

The positions of the waypoints and the vessel's heading and speed that are to be used for each track section are specified by the operator and stored in waypoint tables. Waypoints can be inserted, modified and deleted as required. The vessel's heading is controlled by the following functions:
– Present Heading
– Set Heading
– System Selected Heading

The speed of the vessel along each section of the track can either be taken from the waypoint table or specified on-line by the operator using the Set Vessel Speed function.

Depending on the thrusters' installation and the vessel design, the maximum speed for a vessel in Auto Track low speed mode should not exceed approximately 3 knots since the effect of the lateral thrusters is reduced considerably at speeds higher than this.

The figure 1 shows the track a vessel will follow in Auto Track low speed mode according to the information contained in the table:

Table 1. Autotrack parameters

Waypoint No.	North/East Co-ordinates [m]	Speed [m/s]	Heading [°]
1	1501530 / 503600	0.3	0
2	1501570 / 503680	0.4	330
3	1501650 / 503630	0.4	300
4	1501740 / 503900	0.3	300

The operator can select between two alternative strategies for passing waypoints:
– Slowing down at each waypoint before continuing to the next (used when the vessel must remain on track, even during sharp turns)
– Passing the waypoint at a constant speed on a segment of a circle. The circle's radius can be calculated automatically according to the vessel speed, the angle of turn and the vessel's turning characteristics.

Both strategies generally require solving of the path following problem to acquire the desired dynamical behaviour of the vessel depending on the resultant thrusters' set points (geometric and dynamic problem).

DP vessels have different types of thrusters – tunnel thrusters, azimuth thrusters, main prop and rudders are the most commend used. Both RPM and pitch controlled. The pitch follow-up control is performed from the thruster process station. Set points are received cyclically from the DP controller via the dual communication network or by use of analogue signals. A PID-controller compares actual pitch against set value and controls the hydraulic pitch control valve accordingly.

It would then be convenient to parameterize all path waypoints in terms of a continuous path and constraining the vessel to this path [Skjetne, 2005]. The heading of the vessel could be taken as the direction of the tangent vector along the path, or simply as a constant heading, usually pointed against the environmental forces like waves and wind (system selected heading), or set by the DP operator according to operational demands (for instance to keep the transducer in the range of HPR transponders).

The desired dynamic behaviour along this path would be in first strategy zero speed (fixed positioning) at the waypoints, and when moving along the path from one waypoint to the next the desired path speed should be commanded online by the operator.

In the second strategy the desired dynamic behaviour along the path would be constant speed.

In the works done so far [Skjetne, 2005] the two methods were used: 1) starting with an already available tracking controller, and then converting this into a manoeuvre regulation controller; 2) starting with a parameterized path and a dynamic assignment along the path, designing the control, and tying together the geometric and dynamic objectives with the final pick of an update law. The second method seems to be more flexible and has the advantage that the path variable can be a dynamic state integrated online in the controller to satisfy the dynamic assignment.

These methods will be shortly presented in the following sections and together with the foundations of the fuzzy logic controller combining both.

2 CLASSIC TRACKING CONTROLLER IN DP SYSTEMS

The main functions to be performed in order for a dynamic positioning system to control a given vessel position (x, y) and heading (ψ) are [Cadet, 2003]:
– Estimate vessel motion
– Measure vessel response
– Determine error between prediction and measurement
– Determine corrective action to be applied
– Calculate and allocate appropriate command to thrusters to achieve desired corrective action

Figure 2 presents block diagram of a DP system. The kernel of this system is the simplified hydrodynamic vessel model. This model is a set of equations of motion that is used to predict the motion of the vessel when known forces and moment are applied. In order to separate the wave induced oscillatory part of the motion from the remaining part of the motion, the total vessel motion is modelled as the added outputs of a low-frequency model (LF-model) and a high-frequency model (HF-model). The HF-model represents oscillatory wave components in the vessel motion. The LF-model represents motions induced by wind, thrust and current in surge, sway and yaw. The low frequency portion of the model is controllable by means of thrusters. The algorithm calculates values of vessel's state vector (position, heading and motion variables) by measurements filtration and then it changes resulting force demand - thrusters allocation to meet position, heading and motion settings [Zalewski, 2010].

Thrusters Allocation block is usually ship specific PID controller responsible for achieving dynamic and geometric objectives formulated by the Carrot Computation block. In a Dynamic Positioning application a Kalman filter is used to estimate the state of the vessel (for which a dynamics model has been developed) based on noisy measurements from reference systems and sensors.

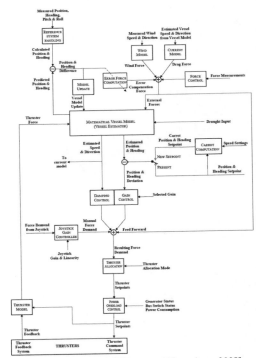

Figure 2. Block diagram of DP system. [Kongsberg, 2008]

2.1 Kalman filtering

The estimation problem solved by the Kalman filter can be expressed as follows [Cadet, 2003]: how to optimally estimate the state of a vessel with an approximate knowledge of the vessel dynamics (imperfect mathematical model) and with noisy measurements from sensors and position reference systems? What is the best state estimate one can get out of all that?

A discrete random dynamic system is described by two equations in the Kalman filter [Zalewski, 2010]:
– state equation (structural model of the process):

$$x_k = A_{k-1} \cdot x_{k-1} + w_k \qquad (1)$$

– measurement equation (measurement model):

$$z_k = H_k \cdot x_k + v_k \qquad (2)$$

where: x – n^{th} dimension state vector,
$\quad w$ – r^{th} dimension state disturbance vector,
$\quad z$ – m^{th} dimension measurement vector,
$\quad v$ – p^{th} dimension state disturbance vector (measurement noise),
$\quad A$ – $n \times n$ dimension transition matrix,
$\quad H$ – $m \times n$ dimension measurement matrix,
$\quad r \leq n, p \leq m.$

Besides, it is assumed for disturbance vectors w and v that they are Gaussian noise of normal distribution, of zero mean vector and are mutually not correlated. The state equation describes the trend of the vector concerned, and the measurement model gives the functional dependence of the measurement on this vector. The solution to the equation system (1), (2), taking into consideration the limitations imposed on disturbance vectors, can be reached by the Kalman filter.

The estimation of the state vector in the filter can be presented by the equations below:
– state vector forecast:

$$\hat{x}_k^- = A_k \hat{x}_{k-1}^+ \qquad (3)$$

where: $\hat{x}_k^- \in \Re^n$ – forecast or a-priori estimated value of the state vector at k-moment, $\hat{x}_k^+ \in \Re^n$ – a-posteriori estimated value of the state vector,
– covariance value of the forecast of state vector:

$$P_k^- = A_k P_{k-1}^+ A_k^T + Q_k \qquad (4)$$

where Q is the matrix of the covariance of the state disturbance (of vector w), index T means matrix transposing,
– filter amplification matrix:

$$K_k = P_k^- H_k^T \left[H_k P_k^- H_k^T + R_k \right]^{-1} \qquad (5)$$

where R is the covariance matrix of measurement disturbance (of vector v),

– estimate of the state vector from filtration $\hat{x}_k^+$ after making measurement z_k:

$$\hat{x}_k^+ = \hat{x}_k^- + K_k \left[z_k - H_k \hat{x}_k^- \right] \qquad (6)$$

– covariance matrix of the estimated state vector:

$$P_k^+ = \left[I - K_k H_k \right] P_k^- \qquad (7)$$

where I is the identity matrix.

In DP systems controlling vessel's position with fixed heading in two dimensions generally the following values are to be estimated: position coordinates (ϕ, λ), projections of the speed vector in relation to the bottom onto the meridian (N-S axis) and the parallel (E-W axis) (V_N, V_E), acceleration vector projections onto the meridian and the parallel (a_N, a_E) and the projections of acceleration vector derivatives in relation to the bottom onto the meridian and the parallel (a'_N, a'_E). In this case the state vector will have the following elements:

$$x_k = \left[\varphi, \lambda, V_N, V_E, a_N, a_E, a_N', a_E' \right]^T \qquad (8)$$

The measured values will be: position coordinates of the positioning system used (ϕ_{PS}, λ_{PS}), speed components in relation to the meridian and the parallel from log or DR navigation performed by the positioning system used (position changes derivatives V_N, V_E), acceleration components in relation to the meridian and the parallel from an inertial transformer MRU (a_N, a_E). So the measurement vector will have the following elements:

$$z_k = \left[\varphi_{PS}, \lambda_{PS}, V_N, V_E, a_N, a_E \right]^T \qquad (9)$$

Some assumptions like values of variance and covariance for particular measurements have to be done, for instance: DGPS system – $\sigma_\phi = 1.0$ m; $\sigma_\lambda = 1.0$ m; coordinates not correlated; speed components – $\sigma_V = 0.1$ m/s; acceleration components – $\sigma_a = 0.01$ m/s^2.

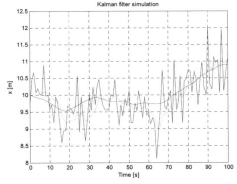

Figure 3. Kalman filter simulation of vessel x co-ordinate estimate.

In the figure 3 the results of the Kalman filter work referring to one co-ordinate are presented (simulated in Matlab®). Blue are measurements, red are unknown true values, green are estimated values.

Based on the simplified vessel model, and using the previous position estimate of the vessel, the prediction step of the Kalman filter gives us a prediction of the vessel position. Based on the forces acting on the vessel, on the vessel model and on the previous position estimate, this is where the DP system thinks the vessel is.

3 PARAMETERIZED PATH TRACKING CONTROLLER IN DP SYSTEMS

As stated in the section 1, the DP Auto Track involves two tasks called the geometric task and the dynamic task. While the main concern is to satisfy the geometric task, the dynamic task, further specified as a speed assignment, ensures that the system output follows the path with the desired speed. The main ideas were published in [Skjetne, 2005].

3.1 *Parameterization of path from waypoints – geometric assignment*

One method to generate a path of class C^r (C^r means that path function f is continuously differentiable up to order r) is to first specify a set of $n+1$ waypoints, and then construct a sufficiently differentiable curve that goes through the waypoints by using splines and interpolations techniques.

Designating the desired path as $p_d(\theta)$ we get n curves $p_{d,i}(\theta)$, $i = 1, 2, \ldots n$, between the waypoints. Each of these is expressed as a polynomial in θ of certain order. Then the expressions for the sub paths are concatenated at the waypoints to assemble the full path. To ensure that the overall path is sufficiently differentiable at the way-points, the order of the polynomials must be sufficiently high. We will consider the path in two dimension space $\Re^2$. Let $l = \{1, 2, \ldots n\}$ be a set of indices identifying each sub path. A column vector will be stated as col. The overall desired curve is denoted $p_d(\theta) = \mathrm{col}(x_d(\theta), y_d(\theta))$, $\theta \in [0, n]$, while individual segments as $p_{d,i}(\theta) = \mathrm{col}(x_{d,i}(\theta), y_{d,i}(\theta))$, $i \in l$, and $p_i = \mathrm{col}(x_i, y_i)$, $i \in l \cup \{n+1\}$ are the waypoints.

Conveniently it is to assume that θ will reach an integer value at each waypoint, starting with $\theta = 0$ at WP 1 and $\theta = i$ at WP i. The differentiability requirement $p_d(\theta) \in C^r$, means that at the connection of two sub paths, the following conditions must hold:

$$\lim_{\theta \to i-1} x_{d,i-1}(\theta) = \lim_{\theta \to i-1} x_{d,i}(\theta) \quad \lim_{\theta \to i-1} y_{d,i-1}(\theta) = \lim_{\theta \to i-1} y_{d,i}(\theta)$$

$$\lim_{\theta \to i-1} x^{\theta}_{d,i-1}(\theta) = \lim_{\theta \to i-1} x^{\theta}_{d,i}(\theta) \quad \lim_{\theta \to i-1} y^{\theta}_{d,i-1}(\theta) = \lim_{\theta \to i-1} y^{\theta}_{d,i}(\theta) \quad (10)$$

$$\vdots \qquad\qquad\qquad \vdots$$

$$\lim_{\theta \to i-1} x^{\theta^r}_{d,i-1}(\theta) = \lim_{\theta \to i-1} x^{\theta^r}_{d,i}(\theta) \quad \lim_{\theta \to i-1} y^{\theta^r}_{d,i-1}(\theta) = \lim_{\theta \to i-1} y^{\theta^r}_{d,i}(\theta)$$

where: $x^{\theta^r}_{d,i} = \dfrac{\partial^r x_{d,i}}{\partial \theta^r}$, $y^{\theta^r}_{d,i} = \dfrac{\partial^r y_{d,i}}{\partial \theta^r}$ for $i \in l \setminus \{1\}$.

The task solution is to determine the coefficients $\{a_{j,i}, b_{j,i}\}$ of k order polynomials:

$$x_{d,i}(\theta) = a_{k,i}\theta^k + \ldots + a_{1,i}\theta + a_{0,i}$$

$$y_{d,i}(\theta) = b_{k,i}\theta^k + \ldots + b_{1,i}\theta + b_{0,i} \quad (11)$$

According to equations (11) for each sub path there are $2 \cdot (k+1)$ unknowns ($(k+1)$ unknowns a_i and $(k+1)$ unknowns b_i) so there are $2n(k+1)$ unknown coefficients in total to be determined for the full path. Two methods are usually used for calculating these coefficients. After linearization of equations set (11) we will get $2n(k+1)$ linear equations $A\phi = c$ for the full path, which can be solved by a matrix inversion operation: $\phi = A^{-1}c$. However, as the number n sub paths increases, this soon encounters numerical problems in the inversion of A. Instead, it is possible to calculate the coefficients for each sub path independently. To ensure the desired continuity at the connection points the numerical values which are common for the neighboring sub paths must be defined.

For a k'th order polynomial $x_{d,i}(\theta)$ we have that $x^{\theta^j}_{d,i}(\theta) = 0$ for $j \geq k+1$. Hence, we can form equations from the first k derivatives of $x_{d,i}(\theta)$:

C^0: $2 \cdot 2n$ equations for $i \in l$:

$$x_{d,i}(i-1) = x_i \quad y_{d,i}(i-1) = y_i$$

$$x_{d,i}(i) = x_{i+1} \quad y_{d,i}(i) = y_{i+1} \quad (12)$$

C^1: four equations for the first and last n-waypoint:

$$x^{\theta}_{d,1}(0) = x_2 - x_1 \quad x^{\theta}_{d,1}(0) = y_2 - y_1$$

$$x^{\theta}_{d,n}(n) = x_{n+1} - x_n \quad y^{\theta}_{d,n}(n) = y_{n+1} - y_n \quad (13)$$

$2 \cdot 2(n-1)$ equations for intermediate waypoints:

$$\left.\begin{array}{l} x^{\theta}_{d,i}(i-1) = \lambda(x_{i+1} - x_{i-1}) \\ y^{\theta}_{d,i}(i-1) = \lambda(y_{i+1} - y_{i-1}) \end{array}\right\} \quad i = 2, \ldots, n$$

$$\left.\begin{array}{l} x^{\theta}_{d,i}(i) = \lambda(x_{i+2} - x_i) \\ y^{\theta}_{d,i}(i) = \lambda(y_{i+2} - y_i) \end{array}\right\} \quad i = 1, \ldots, n-1 \quad (14)$$

where $\lambda > 0$ is a design constant. $\lambda = 0,5$ means that the slope at WP i is the average of pointing against WP i-1 and WP i+1; while $\lambda < 0,5$ means the slope is higher and $\lambda > 0,5$ means the slope is lower.

C^j: $2 \cdot 2n$ equations for $i \in l$, if derivatives of order $j \geq 2$ are 0:

$$x_{d,i}^{\theta^j}(i-1)=0 \quad y_{d,i}^{\theta^j}(i-1)=0$$
$$x_{d,i}^{\theta^j}(i)=0 \qquad y_{d,i}^{\theta^j}(i)=0 \tag{15}$$

which gives $2(j+1)\cdot 2n$ equations to solve for $(k+1)\cdot 2n$ unknowns. The path generation problem is now set up as n linear, decoupled sets of equations $A_i\phi_i = b_i$; $i \in l$; where the unknown vector ϕ_i is:

$$\phi_i = \mathrm{col}(\{a_{j,i}\}_{j=k,\dots,0}, \{b_{j,i}\}_{j=k,\dots,0}) \in R^{2(k+1)} \tag{16}$$

and A_i and b_i are formed correspondingly according to the above equations.

3.2 Dynamic assignment

The second task is to satisfy a desired dynamic behaviour along the path. This can be expressed in terms of a time assignment, speed assignment, or acceleration assignment along the path [Skjetne, 2005].

A time assignment means to be at specific points along the path at specific time instants. For a continuous parameterization $p_d(\theta)$ specific values like θ_1, θ_2; etc., must correspond to specific time instants t_1, t_2; etc., dependent through a design function $v_t(.)$ so that $\theta_1 = v_t(t_1)$ and $\theta_2 = v_t(t_2)$.

A speed assignment is to obtain a desired speed along the path. If $p_d(\theta)$ is a continuous parameterization this can be translated into a desired speed for θ. This desired speed may depend on the location along the path given by θ, or it may explicitly depend on time. A natural choice is therefore to express the desired speed for θ as a design function $v_s = (\theta, t)$.

An acceleration assignment is to obtain a desired acceleration along the path. For a continuous parameterization $p_d(\theta)$ this can be expressed by a design function $v_a = (\theta, \theta, t)$ for θ, which may depend on the speed θ along the path in addition to θ and t.

4 FUZZY CONTROL PATH TRACKING

The design of fuzzy controller solving geometric and dynamic tasks presented in section 3 can be similar to the one described in [Zalewski, 2009].

In case of system being a DP ship, steered in Auto Track mode in accordance to the designed path,

$u(n)$ will be vector of thrusters setting, $y(n)$ will be a vector containing six variables defining actual motion in 3-degrees of freedom: P_{xy} - actual position of selected reference point stored as two variables, v_x - actual longitudinal (advance) velocity, v_y - actual transverse (lateral) velocity, ω - actual angular (rotation) velocity, ψ – actual ship's heading; and $S(n)$ will be a vector containing six variables defining required motion in 3-dof: P_{xyr} - required position of reference point stored as two variables, v_{xr} - required advance velocity, v_{yr} - required lateral velocity, ω_r - required rotation velocity, ψ_r - required ship's heading At time n, $y(n)$ and $S(n)$ are used to compute the input variables of the fuzzy controller (effect of thrusters setting on motion): ΔP_{xy} - deviation between required and actual position, Δv_x - difference or deviation between required and actual longitudinal (advance) velocity, Δv_y - difference or deviation between required and actual transverse (lateral) velocity, $\Delta\omega$ - difference or deviation between required and actual angular (rotation) velocity, $\Delta\psi$ – difference or deviation between required and actual ship's heading. So generally the input variables vector can be designated by:

$$e(n) = S(n) - y(n) \tag{17}$$

Input variable scaling factors are used to conveniently manipulate the effective fuzzification on the scaled universes of discourse. The scaled factors used for $e(n)$ vector in presented research are normalization constants of the five mentioned deviations. Assuming the scaling factors for deviations as vector K_e the scaled input vector is:

$$E(n) = K_e e(n) \tag{18}$$

The scaled variables are then fuzzified by input fuzzy sets defined on the scaled universes of discourse: [0,1]. Figure 4 shows five input fuzzy sets for one of the $E(n)$ parameters that are used by the fuzzy controller implemented in Matlab®.

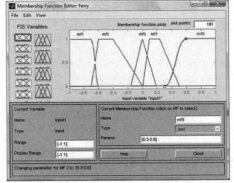

Figure 4. Membership functions of selected input parameter in DP fuzzy logic controller.

The linguistic names "Positive" and "Negative" are related directly to faster speed than required and slower speed than required respectively.

Fuzzification can be formulated mathematically replacing linguistic naming system by a numerical index system, for instance five fuzzy sets used may be represented by A_i, $i = -2$ (NL), -1 (NS), 0 (NZ), 1 (PS), 2 (PL).

No mathematically rigorous formulas or procedures exist to accomplish the design of input fuzzy sets – the proper determination of design parameters is strictly dependent on the experience with system behaviour, hence the expert data coming from ship manoeuvring trials is necessary similarly to designs presented in previous sections.

4.1 Fuzzy rules

Fuzzification results are used by fuzzy logic AND operations in the antecedent of fuzzy rules to make combined membership values for fuzzy inference. An example of a Mamdani fuzzy rule used for control of ship advance speed with main thruster is:

IF $E_2(n)$ is PL AND $E_1(n)$ is NS
THEN $u(n)$ is SAs $\hspace{2em}$ (19)

where PL and NS are input fuzzy sets and SAs (Slow Astern) is an output fuzzy set. In essence rule (19) states that if ship's advance speed is significantly larger than the desired advance speed and the ship's position is a little off the desired track, the controller output should be the pitch setting corresponding to Slow Astern fuzzy set.

The quantity, linguistic names, and membership functions of output fuzzy sets are all design parameters determined by the controller developer. Similarly to input fuzzy sets the most popular membership functions of singleton type have been used.

The exact number of fuzzy rules is determined by the number of input fuzzy sets. For the considered system of ship control the total number of fuzzy rules will be the combination of 5 input variables and 5 fuzzy sets (if for all variables the same number of fuzzy input sets is designed): $5^5=3125$; quite a large amount for only pitches setting. Actually this number of fuzzy rules can be significantly reduced by treating each input variable independently and combining the output during defuzzification. This can be achieved by utilizing coupled fuzzy controllers.

4.2 Fuzzy inference

The resultant membership values of input sets produced by fuzzy logic AND operation [Zadeh, 1996] or product operator can be used [Ying, 2000] are then related to the singleton output fuzzy sets by fuzzy inference. The four common inference methods produce the same inference result if the output fuzzy set is singleton.

If output fuzzy sets in rules are the same fuzzy logic OR operation can be used to combine the memberships.

4.3 Defuzzification

The membership values computed in fuzzy inference must be finally converted into one number by a defuzzifier. In the ongoing research the most prevalent defuzzifier in literature – centroid defuzzifier has been used [Ying, 2000]. The defuzzifier output at time n can be:

$$u(n) = \frac{m_{Z_1} \cdot u_1 + m_{Z_2} \cdot u_2 + m_{Z_3} \cdot u_3 + m_{Z_4} \cdot u_4}{m_{Z_1} + m_{Z_2} + m_{Z_3} + m_{Z_4}} \hspace{1em} (14)$$

where:

$u_1=$-13% of pitch/throttle position (DSAs),
$u_2=$0% of pitch/throttle position (STOP),
$u_3=$-25% of pitch/throttle position (SAs),
$u_4=$-50% of pitch/throttle position (HAs),

$u(n)$ is the new output of the fuzzy controller at time n which will be applied to the ship system to achieve control. In comparison with conventional controllers, what is lacking is the explicit structure of the fuzzy controller behind the presented procedure. On the other hand utilizing expert knowledge for such a black box is much more straightforward and comprehensive.

5 CONCLUSIONS

For many DP operations solving of the path following problem is crucial. It can be achieved by the conventional system modelling methodology, but implementation of the fuzzy controllers seems to be promising as well especially where the system knowledge is represented mostly in an implicit and linguistic form rather than an explicit and analytical form.

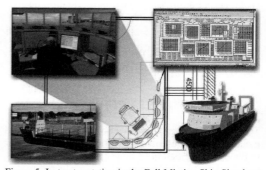

Figure 5. Instructor station in the Full Mission Ship Simulator with DP systems at Maritime University of Szczecin.

REFERENCES

Cadet O., "Introduction to Kalman Filter and its Use in Dynamic Positioning Systems", Dynamic Positioning Conference, Marine Technology Society, September 16-17, 2003.

Kongsberg Maritime AS, "Kongsberg K-Pos DP (OS) Dynamic Positioning System Operator Manual", release 7.1, doc. No. 322281/A, Horten, October 2008

Skjetne R., "The Maneuvering Problem", NTNU, PhD-thesis 2005:1, Trondheim, 2005.

Ying H., "Fuzzy Control and Modelling - Analytical Foundations and Applications", IEEE Press, New York, 2000.

Zadeh L. A., "The evolution of systems analysis and control: a personal perspective", IEEE Control Systems Magazine, 16, 95-98, 1996.

Zalewski P., "Fuzzy Fast Time Simulation Model of Ship's Manoeuvring", Taylor & Francis - Balkema in the Trans-Nav'2009 Proceedings. Gdynia, June 17-19, 2009.

Zalewski P., "Models of DP Systems in Full Mission Ship Simulator", Scientific Journals of Maritime University of Szczecin 20(92), Szczecin, 2010.

14. Simulating Method of Ship's Turning-basins Designing

J. Kornacki
Szczecin Maritime University, Szczecin, Poland

ABSTRACT: The paper presents one of the methods of ship's turning-basins designing. The simulating method is more and more often used to the defining parameters projected turning-basins, testing of existing turning-basins and the improving of the manoeuvring practice on the particular manoeuvring basin.

1 INTRODUCTION

The manoeuvre of ships turning is executed every time during the ships presence in the port and it is one of the often port manoeuvres. The influences on the size of turning basin during the manoeuvre have the large quantity of factors.

The turning basin has two meanings. First meaning is the manoeuvring basin delimited by the manoeuvring ships, second is the hydro-technical building artificial or natural with suitable horizontal and vertical dimensions, where the considerable alterations of the course of the ship are executed. Obviously, the turnings over are practices „in the place". This should be understand as the change of the course of the ship whose linear speeds, during the manoeuvre, are close to zero. Turning the ship over is done on the turning basin as a result of the planned tactics of manoeuvring and can be done on itself or in co-operation with tugs or use of anchors or spring lines. All dimensions of turning basin as the hydro technical building has to be larger than the turning basin understood as the manoeuvre basin to avoid the collision with bottom or bank (Kornacki 2007).

The simulating method of designing the parameters of turning basins are based on series of tests in comparable conditions on prepared model of reservoir and the model of the ship planned to use the turning basin. The results of tests are subjected the statistical processing. Effect of that kind of research is delimitation of the area of manoeuvring on the turning basin according to the various foundations of hydro meteorological conditions, various parameters of ships and various levels of the trust. Characteristic feature of the simulating method is that simulating models of the ship manoeuvring are especially designed to the solved problem.

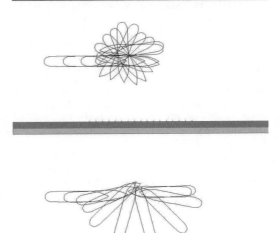

Figure 1. The example of tests of simulating of turning manoeuvres.

The material in the result of simulating investigations comes into very large sizes, which is subjected far processing. The application of the methods which will let process got results to the form enabling making far analysis necessary is. The area of manoeuvring of the ship is the basic criterion of the analysis of the results of simulating testing's, and the dimensions of this area are it numerical coefficient. The variety of the elements of the system of the port and narrow waters causes that various methods are applied. These methods are characterized limitations and conditions. The method of parallel sections, sector method and polar method are complies with marking the dimensions of the basins of manoeuvring of the ship in simulating investigations (Guziewicz & Ślączka, 1997).

2 DELIMITING THE BASIN OF MANOEUVRING OF THE SHIP ON THE TURNING BASIN

The methods of delimitation of the area of manoeuvring of the ship are based on the suitable division of the water areas of manoeuvring and the engagement of suitable point or the axis of the reference. For the needs of delimitation of the manoeuvring area on the turning basin and during the manoeuvres of the turning of the ship the polar method is engaged. This method differs from the method of parallel sections that the sections are replaced by sectors and the axis of the reference is replaced by the point of reference. The difference in the relation to the sector method depends in the engagement of the point of reference not outside but inside the area of manoeuvring and the division of the manoeuvring basin on sectors hugging the round angle.

The polar method in the principle is the change of the sector method. Co-ordinates of points mark the area of manoeuvring ship are qualified in the polar co-ordinates. Then, knowing the exact position of the point of the reference, they are easy to proceeding in the far analysis. Wanting delimit the manoeuvring basin in the polar method, the studied manoeuvring area should split on the sectors of the delimited width $\Delta \alpha$.

The selection of the proper width of the sector essential is. The width of the sector simplifying should fulfil dependence (Guziewicz & Ślączka, 1997):

$$\Delta \alpha \rangle \frac{L_{OA} \times \sin \beta \times 180}{(i-1) \times R \times \pi} \ [^{\circ}] \tag{1}$$

where:
$\Delta \alpha$ - width of the sector [°],
L_{OA} - length over all [m],
β - acute angle contained among the longitudinal axis of the symmetry of the ship and the secant of definite sector [°],
i - the number of extreme points on one board describing the waterline of floatation [-],
R - the ray of the projected turning basin [m],

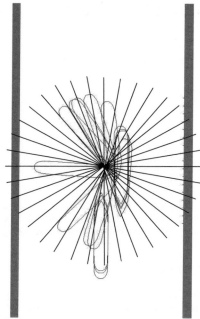

Figure 2. The split of the manoeuvring area in the polar method.

The selection of the width of the sector is dependent from the received suitable number of the extreme points of the ship. This influences on the size the error steps out among the delimited area of the manoeuvre and real manoeuvring area left by the ship.

One can express this error in the approximation (Guziewicz & Ślączka, 1997):

$$\delta_{Si} = \Delta \alpha \times R_i \times \cos \beta \times \frac{\pi}{360} \ [\text{m}] \tag{2}$$

where:
δ_{Si} - error of delimiting manoeuvring basin [m],
$\Delta \alpha$ - width of the sector [°],
R_i - the ray of the projected turning basin [m],
β - acute angle contained among the longitudinal axis of the symmetry of the ship and the secant of definite sector [°],

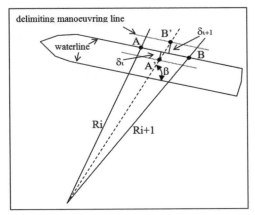

Figure 3. The error of delimiting manoeuvring basin.

During the test data with the course of the ship, shape of the waterline of floatation and the co-ordinates of geometrical centre of the waterline of floatation are recorded. The co-ordinates of extreme points of the ship are calculated in the polar co-ordinates.

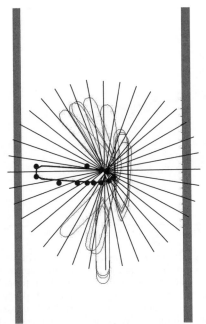

Figure 4. The extreme points of the ship on one particular waterline.

Distances d_i of the extreme points of ship from the point of the reference are calculated, where i is the number of the extreme points of the ship describing the waterline of flotation. The table of the distance D_S [k, s, d_i] in which counted distances d_i are assigned to sectors s for individual simulating tests k is created.

Based on the table of the distance D_S [k, s, d_i] the tables of maximum distances $D_{S\ max}$ [k, s, $d_{i\ max}$] and minimum distances $D_{S\ min}$ [k, s, $d_{i\ min}$] of extreme points of ship waterline from the point of reference in every sector of basin for every simulating test are created. This makes possible the assignment the line of movement of the ship in the single test.

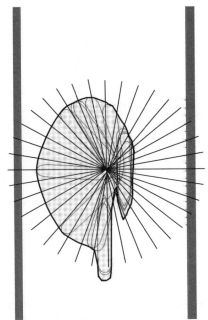

Figure 5. The line of movement of the ship in the single test.

Based on the table of maximum distances $D_{S\ max}$ [k, s, $d_{i\ max}$] and minimum distances $D_{S\ min}$ [k, s, $d_{i\ min}$] and the suitable statistical model of the expansion of maximum and minimum distances, the co-ordinates of points of the area of manoeuvring with the assumption of level of the trust are appointed in separate sectors.

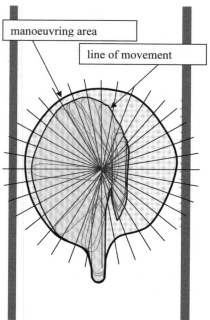

Figure 6. The line of movement and manoeuvring area of the ship.

3 THE PRACTICAL USE OF THE SIMULATING INVESTIGATIONS

3.1 *The practical use of the polar method*

The polar method can be applied to the preparation of the measuring data to the statistical processing during real investigations. Error resulting from the applied method of delimitation the line of movement in this case is enlarged by error resulting from the measurement and error resulting from the preparation of the measuring data to the use of the polar method becomes.

In the case of use of the polar method, while delimiting the manoeuvring basin of ship in simulating investigations, the results are burdened the only error the applied method.

In the practice, the simulating method complies in two aims. First, it complies in the qualification of sizes of the planned turning basin. Second, it complies in the qualification of maximum permissible sizes of ships can safely use the turning basin.

The turning basin is safe for the ships manoeuvring, if every her sizes in the horizontal plane and perpendicular plane are larger than the sizes of the manoeuvring area traced by the manoeuvring ship. It is mean that on the whole area the safe under keel clearance and the safe distance from banks and slopes has to be kept.

The analysis of the results of simulating tests leads to measuring the parameters of manoeuvring area, which means the qualification of the parameters of the horizontal safe manoeuvring area.

Figures 7 and 8 present the vertical and the horizontal section view of turning basin.

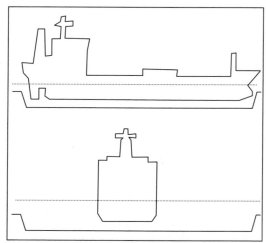

Figure 7. The turning basin – the vertical view.

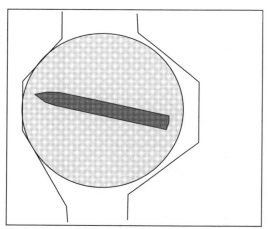

Figure 8. The turning basin – the horizontal view.

During simulating investigation relevant to the turning basin, it is important to take into consideration also the expansion of the speed of propeller streams, the expansion of directions of propeller streams, and influence of the every kind of hydrotechnical buildings, slopes and banks on the ship (Galor, W. 2002).

3.2 The examples of use of the simulating investigations

Figures 9, 10 and 11 present the examples of results of simulating investigation of typical turning manoeuvre.

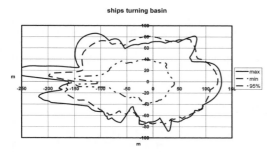

Figure 9. The manoeuvring areas with different trust levels during turning manoeuvres.

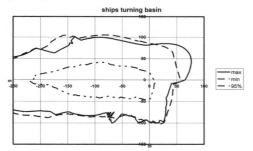

Figure 10. The manoeuvring areas with different trust levels during turning manoeuvres.

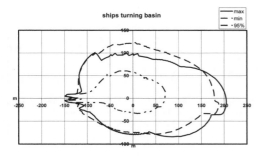

Figure 11. The manoeuvring areas with different trust levels during turning manoeuvres.

Figures 12, 13, 14 and 15 present the examples of results of simulating investigation on real port water areas (Kornacki, J.& Galor, W. 2007).

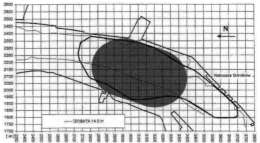

Figure 12. The north turning basin investigation in port of Świnoujście (The unpublished report 2000).

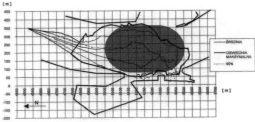

Figure 13. The south (BPP) turning basin investigation in port of Świnoujście (The unpublished report 1995a).

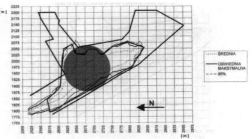

Figure 14. The turning basin investigation in port of Kołobrzeg (The unpublished report 1995b).

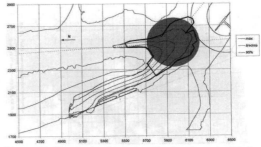

Figure 15. The turning basin investigation in port of Police (The unpublished report 1998).

These examples are showing scale of use of the simulating investigations. Base on the northern turning basin in Świnoujście, the problem of the analysis of the turning basin can be present.

The northern turning basin in Świnoujście was investigated several times, usually in the context of considerations on the subject of the possibility of manoeuvring in the harbour of the "maximum" ships. Problem this came back because of the development of the harbour in Świnoujście and the ambitions of the service larger ships. In the years 1995-97 was made one of first analyses in the aspect of the possible modernization of the northern turning basin. The investigation was made then to qualify possibilities of the entry the ships about the total length 255 m into the harbour. The qualification of the possibility of the safe turn of ships on the northern turning basin was one of the elements of these investigations. The eventuality of the increase of the sizes of the turning basin was taken into the consideration then so that ships do not violate the infrastructure of the western bank of Świna. In the year 2001, the problem was returned when the possibility of entering into the harbour of the ships of total length 260 m were study again. The investigations were led with the assumption that the existing bathymetry of the port's reservoir will not be changed. In the year 2003, the possibility of the entry the ships of total length to 280 m with new tugboat „Uran" was analysed.

4 CONCLUSIONS

The methods of delimiting the area of manoeuvring basin of the ship depend on the suitable split of the manoeuvring area. Polar method should be applied to marking manoeuvring area on turning basin.

The polar method is based on the suitable split of the manoeuvring area, the engagement of the point of reference and proper random variable to assignment of this basin. The distance of points describing the waterline of the ship from the point of the reference is the applied universally random variable.

The taken value of the width of the sector in the polar method has the direct influence on the value of the error among the real line of the movement of the ship and the line of movement of the ship delimited the polar method.

Main advantage of the simulating method is the possibility of the exact qualification of the manoeuvring area. Through the far statistical analysis of many tests, the ready information of the sizes of the basin which the ship needs to the realization of the manoeuvre of the turning is available.

Very important advantage of the simulating method is relatively low cost of investigation.

Often, navigational obstructions mark the accessible manoeuvring basin. It has the influence on the way of manoeuvring. The use of the simulation allows optimizing sizes of such basin or the way of manoeuvring.

REFERENCES

Briggs M.J. &Borgman L.E. & Bratteland E.2003. Probability assessment for deep-draft navigation channel design. International Journal for Coastal Harbour and Offshore Engineering, Coastal Engineering 48, Elsevier.
Galor W. 2002. Bezpieczeństwo żeglugi na akwenach ograniczonych budowlami hydrotechnicznymi. Fundacja Rozwoju Wyższej Szkoły Morskiej w Szczecinie.
Gucma, S. & Jagniszczak, I. 2006. Nawigacja dla kapitanów. Gdańsk: Fundacja Promocji Przemysłu Okrętowego I Gospodarki Morskiej.
Guziewicz, J. & Ślączka, W. 1997. Szczecin: The methods of assigning ship's manoeuvring area applied in simulation research, International Scientific and Technical Conference on Sea Traffic Engineering, Maritime University of Szczecin.
Kornacki, J. & Galor, W. 2007. Gdynia: Analysis of ships turns manoeuvres in port water area. TransNav'07.
Kornacki, J. 2007. The analysis of the methods of ship's turning-basins designing. Szczecin: XII International Scientific and Technical Conference on Marin Traffic Engineering, Maritime University of Szczecin.
The unpublished report 1995a. Badania symulacyjne ruchu statków w bazie paliw płynnych w Świnoujściu. Maritime University of Szczecin.
The unpublished report 1995b, Określenie optymalnego wariantu przebudowy wejścia do portu Kołobrzeg w oparciu o badania symulacyjnego ruchu statków, Maritime University of Szczecin.
The unpublished report 1998, Badanie możliwości optymalnej lokalizacji przeładowni kwasu fosforowego w Porcie Police. Maritime University of Szczecin.
The unpublished report 2000. Analiza nawigacyjna dla przebudowanego Nabrzeża Górników w porcie handlowym Świnoujście. Naval University of Szczecin.

15. Capabilities of Ship Handling Simulators to Simulate Shallow Water, Bank and Canal Effects

L.K. Kobylinski

Foundation for Safety of Navigation and Environment Protection Iława Poland

ABSTRACT: Safe operation of ships in restricted areas, in particular in canals and waterways of restricted width and depth, often with presence of current. depends on operator skill. One way to influence operator skill and hence to increase safety against collisions and groundings is proper training of operators in realistic environment. Training could be accomplished on board ships, which takes, however, long time but also on simulators. There are two types of simulators: full mission bridge simulators (FMBS) working in real time and physical simulators using large manned models in purposeful prepared training areas (MMS). Capabilities of both type simulators are discussed in detail. Capability of FMBS depends on computer codes governing them. Few examples of capability of FMBS to reproduce correctly ship handling situations are shown. There are few MMS in the world, one of which is Ilawa Ship Handling Research and Training Centre. In the centre models of several types of ships are available and training areas are developed representing different naviga-tional situations. The main purpose of the training exercises is to show the trainees how to handle the ship in many close proximity situations, in the presence of current, in very restricted water areas etc.

1 INTRODUCTION

Collisions, rammings and groundings, so called CRG casualties, constitute large part of all casualties at sea (approximately around 60 per cent) (Samuelides & Friese 1984). Therefore reducing risk of CRG casualties contributes largely to the reduction of overall risk of sea voyage.

Risk of CRG casualty depends on several factors, one of which is human factor, i.e. operators skill. Published analyses associated with commercial shipping during recent years indicated that human errors that occurred during handling operations were responsible for approximately 62 per cent of the major claims figure (Payer 1994). Other sources show, that about 80 % of all CRG casualties are results of human failure. Therefore attention is focused recently to the role of human factor in safety. (US Coast Guard 1995).

As about two thirds of all CRG casualties are caused by human error it is necessary to analyse factors which contribute to the efficiency of the operator. The author discussed this subject in the paper presented to Nav 2009 (Kobylinski 2009) showing that one of the most important factors contributing to this is training.

2 SIMULATOR TRAINING

There are several factors contributing to the reduction of the number of CRG accidents, and experience is one of them. Experience is gained over years of practice. Specialized training on simulators accelerates gaining experience, in particular gaining experience in handling dangerous situations that may be rarely met during operation of real ships. Therefore specialized training in ship handling is required by the International Maritime Organisation. Seafarers' Training, Certification and Watchkeeping (STCW) Code, Part A, includes mandatory standards regarding provisions of the Annex to the STCW Convention. Apart training onboard ships, approved simulator training or training on manned reduced scale ship models is mentioned there, as a method of demonstrating competence in ship manoeuvring and handling for officers in charge of navigational watch and ship masters.

Also ship owners companies and pilots organizations attach recently great importance to training on simulators and some pilots organizations require repetition of such training every 5 years.

Obviously the best way to train ship officers and pilots in shiphandling and manoeuvring is to perform training onboard real ships. Any use of simulators should be in addition to training onboard ships. However, gaining skill "on job" watching experienced practitioner working is a long and tedious process. Moreover certain handling situations including some critical ones may never occur during the training period onboard ships and no experience how to deal with such situations could be gained this way. When serving on ships engaged in regular service there is little or no possibility to learn about handling in critical situations because such situations must be avoided as far possible.

Simulator training is expensive, therefore the simulator courses must utilize time available in the most effective way. In order to achieve positive results simulators must be properly arranged and the programme of simulator exercised should be properly planned in order to achieve prescribed goals.

In general, simulators may be either equipment or situations. A simulator is defined as any system used as a representation of real working conditions to enable trainees to acquire and practice skills, knowledge and attitudes. A simulator is thus characterised by the following:

– imitation of a real situation and/or equipment which, however, may permit, for training purposes, the deliberate omission of some aspects of the equipment in operation being simulated, and
– user capability to control aspects of the operation being simulated.

The effectiveness of a simulator in training mariners depends on the simulator capabilities to simulate the reality. Sorensen (2006) stressed the point that simulators must be realistic and accurate in simulating the reality. Therefore simulators should, apart from simulating properly the main manoeuvring characteristics of a given ship, i.e.

– Turning characteristics
– Yaw control characteristics
– Course keeping characteristics and
– Stopping characteristics

be capable to simulate different factors influencing ship behaviour, e.g: at least:

– Shallow water effect
– Bank effect
– Effect of proximity of quay or pier
– Effect of limitation of dimensions of harbour basin
– Surface and submerged channel effect
– Ship-to-ship interaction
– Effect of current
– Effect of special rudder installations, including thrusters
– Effect of soft bottom and mud
– Ship-tug cooperation in harbour (low speed towing) and.
– Escorting operations using tugs
– Anchoring operations.

3 FULL MISSION BRIDGE AND MANNED MODELS SIMULATORS

Simulators used in training in ship handling and manoeuvring are basically of two types : Full Mission Bridge Simulators (FMBS) and Manned Models Simulators (MMS).

FMBS computer controlled simulators are widely used for training of ship officers, pilots and students of marine schools and also for studying various manoeuvring problems, first of all problems associated with the design of ports and harbours.

There is at present a considerable number of such simulators of different types operating throughout the world, starting from desk simulators to sophisticated FMBS where the trainee is placed inside a bridge mock-up with actual bridge equipment, realistic visual scene of the environment, and sometimes rolling and pitching motions and engine noise.

FBMS are working in the real time and are controlled by computers programmed to simulate ship motion controlled by rudder and engine (and thrusters or tugs) in different environmental conditions

MMS use large models for training purposes in specially arranged water areas, ponds or lakes. Models are sufficiently large in order to accommodate 2-4 people (students and instructors) and are constructed according to laws of similitude. Models are controlled by the helmsman and are manoeuvring in the areas where mock-up of ports and harbours, locks, canals, bridges piers and quays, shallow water areas and other facilities are constructed and where also routes marked by leading marks or lights (for night exercises) are laid out all in the same reduced scale as the models. Also in certain areas current is generated. As a rule, monitoring system allowing to monitor track of the model is available.

Important feature of manned model exercises is that all manoeuvres are performed not in real time, but in model time which is accelerated by the factor λ^{-1}. This may pose some difficulties for trainees at the beginning who must adjust to the accelerated time scale.

Currently there are only few training centres using manned models in the world, however, according to the recent information, few others are planned or even in the development stage.

4 CAPABILITIES OF FBM SIMULATORS

In FMBS because there is a mathematical model of ship motion on which computer codes are based it is important that this mathematical model represents properly behaviour of the real ship. In spite of great progress in the development of the theoretical basis of ship manoeuvrability not only in unrestricted water areas (turning, course-keeping and stopping characteristics), but also in the proximity of other objects (bank, shallow water effects and the effect of other ships), the last effects are still investigated not sufficiently enough. Sophisticated computer programmes that include calculations of hydrodynamic coefficients using advanced methods requiring powerful computers and extreme large memory. simulating the close proximity effects cannot be used in FBMS because they must work "on line" therefore rather simplified methods must be developed for this purpose.

Practically all modern FMBS are capable to simulate manoeuvring and ship handling characteristics in open water properly. Usually they are also capable to simulate the close proximity effects based on simplified theory. But in many cases even simple manoeuvres such as turning circle manoeuvre or zigzag manoeuvre are often simulated not accurately enough. Gofman & Manin (1999, 2000) showed several cases where results of simulation on Norcontrol SH simulator differed considerably from results obtained during tests of full-scale ships. One may however argue that results shown by Gofman were obtained in nineties of the last century and modern simulators are much more effective.

There is little information available on the validation of the effectiveness of FBMS. Some data on comparison of simulated and measured at full scale trials of few ships were collated by Ankudinov (2010) and one example of simulation of turning circle test on TRANSAS simulator is shown in Table 1.

Fig 1 shows results of comparison of simulated and measured characteristics of stopping manoeuvre of the ship ARKONA. Simulator in this case was ANS 5000 developed by Rheinmetall Defence Electronics GmbH (de Mello Petey 2008). In both cases it is seen that the simulation is quite reliable.

Results of simulation of manoeuvring capabilities of POD driven ships on this simulator are also available (de Mello Petey 2008) and by Heinke(2004). The code used in this simulator takes into account the following:
– Propeller thrust
– Transverse propeller force
– Lift and drag forces of the POD body
– Interaction effects between different POD units
– Interaction effects between POD and hull, and
– Shallow water effects.

Table 1 Turning circle tests with both pods at an angle 35⁰ (EUROPA)

	Manoeuvre to port		Manoeuvre to starboard	
	Simu-lated	Actual	Simu-lated	Actual
Starting speed [knots]	21.40		11.40	
Engine[%]	100		60	
Rudder angle [deg]	35.0		-35.0	
Adcance [m]	404.0	379.6	333.0	364.0
Transfer [m]	165.0	159.1	167.0	164.3
Tactical diameter [m]	375.0	392.1	382.5	398.7
Turning circle diameter [m]	320.0	313.7	323.5	320.3
Steady speed at turn [knots]	6.40	6.59	3.90	4.38
t90 [s]	56	54	91	96
t180 [s]	117	120	182	203
t270 [s]		192		314
t360 [s]	260	264	397	425

The high level of accuracy achieved by the simulation module was proved by validation tests performed with pollution control ship ARKONA (L= 69.2m). The example of comparison of simulated and measured results of the stopping manoeuvre where at full speed both POD were commanded to zero RPM is shown in fig. 1 (de Mello Petey 2008).

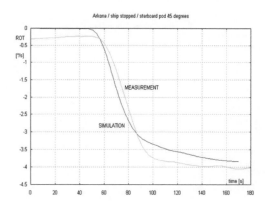

Figure 1. Comparison of simulated and measured characteristics of stopping manoeuvre ARKONA ship (Ref. 27)

The technique used by TRANSAS in simulating manoeuvring characteristics of ships in shallow water and the bank effect is based on the generalized flow pressure functional describing motion effects and variable pressure field of maneuvering ship in the restricted channel of variable bottom and banks in the presence of other stationary or moving ships. The developed technique is fairly complex and best suited for solid unmovable objects in the channel (walls, moored ships). The modeling of proximity of other maneuvering ships of various types moving with various heading angles and velocities needs however further refinement (Ankudinov 2010).

Gronarz (2010) reported results of the simulation of shallow water and bank effect in four most modern FBMS, marked A,B,C,D. The results of simulation of speed loss in shallow water and increase of turning diameter are shown in figs. 3 and 4.

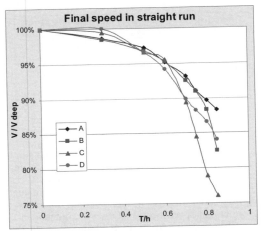

Figure 2. Speed loss with reduced UKC

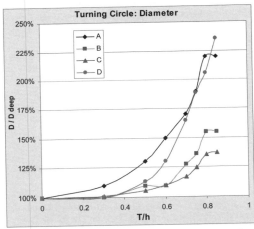

Fig 3 Increase of turning circle diameter with reduced UKC

In deep water a ship can reach the highest velocity using constant revolutions of the propeller. With reduced UKC, i.e. increased T/h the speed loss will increase. In general all simulators show loss of speed with decreasing depth of the water as it can be seen in fig. 2. However, for T/h=0.3 i.e. the case where UKC is more than twice the draught of the ships gap the speed loss should only be marginal. This is represented correct for simulators C and D, but simulators A and B show significant loss of speed which is not correct. On the other hand at very shallow water

(T/H~0.8) the speed loss shown by simulators A and B is not great enough as it should be.

The simulated turning circle diameters in shallow water are larger, as expected (Fig.3).

However simulator A shows increase of turning diameter in rather deep water (T/H=0.3) which is not correct. As it is seen from fig.3 the increase in diameter in shallow water is significantly different. The range of 35% (C) to 135% (D) increase seems unusual. Also for simulators A, B and C the turning diameters are nearly constant for T/h=0.80 and 0.85 which contradicts the theory. This means that simulation of shallow water effects are not represented by all simulators correctly and the computer codes used have to be improved.

5 CAPABILITIES OF MMS SIMULATORS

In the case of manned models the governing law of similitude is Froude's law and all quantities for models are calculated according to the requirements of this law. However, as it is well known, the requirements of second law of similitude which is relevant to ship motion, Reynolds law, cannot be met. This means that the flow around the ship hull and appendages and in particular separation phenomena might be not reproduced correctly in the model scale. Fortunately those effects are important when the models are small. With models 8 to 15 m long the Reynolds number is sufficiently high to avoid the majority of such effects.

One important difficulty with manned models is impossibility to reproduce wind effect. Wind is a natural phenomenon and according to laws of similitude wind force should be reduced by factor λ^3 (λ - model scale). Wind force is proportional to the windage area and to the wind velocity squared. Windage area is reduced automatically by factor λ^2 but wind velocity apparently cannot be reduced. However, actually windage area in models is usually reduced more than by factor λ^2, and wind velocity due to sheltered training area and low position of the windage area in the model in comparison with the full-scale ship is considerably reduced. Still usually wind force is larger than it should be.

Capability of manned models to simulate shallow water, bank, submerged and surface canal effects, effect of current, close proximity of other stationary or moving objects is automatically assured and is practically unlimited, restricted only by local conditions in the training area.

As there are only few manned model centres operating in the world, facilities arranged in the Iława Ship Handling Research and Training centre are shown below as an example.

As safe handling of ships is much more difficult in restricted areas and in presence of the current, in Iława Ship Handling Centre there are artificially

prepared training areas that, apart of the standard model routes marked by leading marks, leading lights (at night) and buoys, comprise also routes particularly suitable for training ship handling in canals and shallow and restricted areas. They include:
- restricted cross-section surface canal of the length 140m (corresponding to 3.3 km in reality), called Pilot's Canal. In this canal exercises comprising passing the canal feeling bank and restricted cross section effects, stopping ships in restricted width of the fairway, meeting and overtaking with two or three ships feeling interaction effects are performed,
- wide (corresponding to about 360m width in reality) shallow water canal of the length corresponding to about 1.5 km, where current could be generated from both sides, called Chief's Canal. Passing the shoal, feeling slowing down and squat, berthing in shallow water, turning the ship in shallow water and in current and similar exercises are performed.
- long (corresponding to about 2.5 km in reality), narrow deep water waterway comprising several bends, marked by buoys, simulating some routes in fiords and similar areas called Captain's Canal,
- narrow fairway restricted from one side by the shore, called Bank Effect Route where ships are supposed to feel bank effect,
- narrow passages, including narrow passage under the bridge feeling the close proximity effect,
- river estuary area where several current generators installed create current. Several mooring places are provided in the estuary, including sheltered dock. Current pattern and velocities could be adjusted by activating particular current generators, the maximum current velocity correspond to 4 knots in full scale.(fig.4). There is possibility to arrange several exercises where ships make manoeuvres in current.
- locks, deep and shallow water docks for docking ships in different situations, harbour basins of different dimensions and configuration of the entrance
- mock-up harbour basins, locks, bridges, fairways and other arrangements existing in different parts of the world as the need arises.

6 CONCLUSIONS

It appears that simulator training in ship handling becomes more and more popular and some pilots organizations require now refreshing such courses every five years.

From the experience with FBMS and MMS simulators it is now clear that they do not supersede but rather they supplement each other, because the purpose of training on each of them is different.

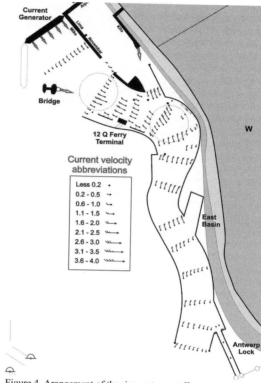

Figure 4. Arangement of the river estuary at Ilawa centre

Capability of FBMS depends on the reliability of computer codes governing them, that are still far from perfection, and the quality of visualization of the situation around the ship simulated. They are particularly suitable to simulate situations in some ports, canals, approaches etc, and master and pilots may learn how to maneuver in this particular situation. FBMS may be used also as a tool for harbour design.

The capability of MMS depends on the possibility of making different arrangements such as described above in the training area available. From this point of view Ilawa training centre in comparison to the other centres (Port Revel and Warsash) has the advantage of having to its disposal large water area (Silm lake),where different arrangements could be installed.

The purpose of training on manned models is mainly to make the trainees aware of different hydrodynamic effects, in particular close proximity interactions, which may be easily arranged. High realism and automatically hydrodynamic correct representation of close proximity situations is the main advantage of MMS.

Tugs action, escort and anchoring and ship-to-ship operations, simulation of which is attempted also in FBMS are particularly realistic when using manned models.

REFERENCES

Ankudinov V.(2010): Azipod cruise ship. Manoeuvring in deep and shallow water. AZIPILOT Project Report WP2

de Mello Petey F. (2008): Advanced podded drive simulation for marine training and research. International Marine Safety Forum Meeting, Warnemuende

Gofman A.D., Manin V.M.(1999): Ship handling simulators validity - the real state and the ways of mathematical models correction. International conference. HYDRONAV'99 - MANOEUVRABILITY'99, Ostroda.

Gofman A.D., Manin V.M.(2000): Shiphandling simulator validity. Validation and correction of mathematical models. International Conference on Marine Simulation and Ship Manoeuvring, MARSIM, Orlando.

Gronarz A. (2010): Shallow water, bank effect and canal interaction. AZIPILOT Project Report Wp 2.2a

Heinke H.J. (2004): Investigations about the forces and moments at podded drives. 1st International Conference on Technological Advances in Podded Propulsion, Newcastle

Kobylinski L. (2008): Training for safe operation of ships in canals and waterways Proceedings SOCW Conference, Glasgow

Kobylinski L. (2009):Risk analysis and human factor in prevention of CRG casualties Marine Navigation and Safety of Sea Transportation. A. Weintrit editor CRC Press

Payer H. (1994): Schiffssicherheit und das menschliche Versagen. Hansa-Schffahrt-Schiffbau-Hafen, 131 Jahrgang, Nr.10

Samuelides E., Friese P. (1984): Experimental and numerical simulation of ship collisions. Proc. 3rd Congress on Marine Technology, IMAEM, Athens

Sorensen P.K (2006): Tug simulation training - request for realism and accuracy. International Conference on Marine Simulation and Ship Manoeuvring, MARSIM 2006,

U.S.Coast Guard (1995): Prevention through people. Quality Action Team Report.

16. Development of a Costs Simulator to Assess New Maritime Trade Routes

F.X. Martínez de Osés, M. Castells i Sanabra & M. Rodríguez Nuevo
Nautical Science and Engineering Department, Universitat Politecnica de Catalunya, Barcelona, Spain

ABSTRACT: This paper is going to describe the design process of a simulator that assesses the costs of different means of transport. The evaluation not only will be done regarding the internal costs but also the external costs that will be translated to environmental costs, based on existing databases. The paper shows the development carried out to create this simulator and analyse all components of the logistical chain, i.e. port operation costs, road haulage costs and maritime leg costs. The simulation results have been validated with real data of actual maritime routes to check its reliability. As a conclusion, the costs simulator permits assess costs of new maritime trade routes comparing them with road transport.

1 INTRODUCTION

One of the main challenges identified by the European Commission White Paper on transport was to address the imbalance in the development of the different modes of transport. Specific actions looking to boost rail and maritime connections were foreseen and included the establishment of the Marco Polo programs. The demand for increased mobility and increased flexibility and timeliness of delivery has led to road transport becoming the dominant mode of transport in the European Union. The growth in road transport has had a significant impact on road congestion, road safety, pollution and land use. In view of this, a change from traditional unimodal to multimodal transport is desirable. Maritime transport is one of the least pollutant modes. Additionally, it contributes to the reduction of traffic congestion, accidents and noise costs on European roadways. Another advantage of ships over trucks and trains is that vessels consume less fuel as a result of the relatively low speeds at which they travel. All these advantages justify support actions to intermodal chains with marine sections including Short Sea Shipping (SSS) as a way to reach more sustainable mobility within Europe. The main benefit of Short Sea Shipping lies in the possibility of combining the inherent advantages provided by the involved modes, thus reducing costs and increasing freight transport capacity over long distances.

Nevertheless, maritime society still regards maritime transport as a slow, inefficient mode since shippers do not yet offer the best value for money.

This paper shows the development carried out to create a simulator that compares freight transport by only road chains and by multimodal (with SSS) chains and comprises five sections. The introductory section summarises the context of the paper. Section two describes the actual scenario related on the state of art of freight costs simulators. A brief description of the methodology used in the development of the costs simulator is explained in section three. The results obtained using costs simulator tool are shown in the next section. Finally conclusions and further research are drawn in section five.

2 FREIGHT COSTS STATE OF ART

The determination of cost functions and variables is important in assessing the feasibility of a process. Historically, the empirical estimation of port cost functions started in the 60's with Wanhill's work. The works by De Monie, Dowd and Lechines, Talley and Conforti proposed a cost analysis to appraise port performance and output by calculations of several indicators. Ametller Malfaz thesis describes the development of cost and time evaluation under the hypothesis of freight distribution based on population density. Actual cost and time simulators an be divided into two groups: the first one a few parameters must be introduced to determine cost or time without specifying the method used; the second one calculates external costs based on theoretical studies like Realise and Recordit projects.

3 DEVELOPMENT OF THE COST SIMULATOR

This paper presents a simulator of internal and external costs which also allows updating data. In order to design the simulation model, the behavior of freight distribution systems must be known. Road haulage, port operation and maritime leg must be modeled to assign costs derived from each of the parts or components of the logistic chain.

The calculation method of overall costs (in €) have been calculated considering a single variable (Gross Tonnage) for all transport modes.

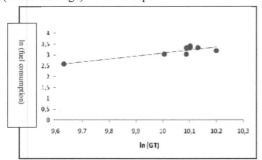

Figure 1.Example of logarithmical ratio between fuel costs and gross tonnage (GT). Source: Own

Time (in hours) spent by the modes of transport to move freight between an origin and a destination will strongly depend upon the physical and operation speed of the modes employed. Calculations consider European road transport regulations on driving times and costs of road freight transport.

Data of truck internal costs will be obtained by analyzing a set of model trucks specified by the Ministry of Public Works and for vessels, Short Sea Shipping Ro-Ro ships are ships employed in Mediterranean maritime routes.

All required data is computed by an engine generated by an Excel spreadsheet and a computer program complied in Visual Basic, and then presented in tables and graphs.

After we have introduced all required data, the methodology used by de simulator can be summarized by the following steps:

1 Choose data from the destination matrix and find out whether there is a destination for the selected route.
2 Choose data from the origin matrix and find out whether there is an origin for the selected route.
3 Choose data from the maritime distance matrix.
4 Introduce ship occupancy rate (σ).
5 Introduce type of freight (σ).
6 Introduce the number of stops made by the ship in each trip (ρ).
7 Introduce specific company profits as a payment for ship services (β).

8 Print and display all solutions for the selected ship (calculation of Short Sea Shipping and only road transport costs and pollutant emission costs).
9 Choose the three best ships for the selected route from the simulator's database and provide their particulars (ship's name, year of building, length, breath, tonnage, lane meters, power, speed and number of platforms).
10 Perform routines under the established formulation.

All data are interpreted by means of tables, charts and mask designed for the presentation of simulator data.

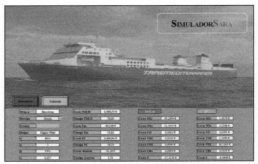

Figure 2. Example of the mask showing parameters to be selected before the calculation process and results of external and internal costs. Source: Own

4 PRELIMINARY RESULTS EXAMPLE FROM THE COSTS SIMULATOR

Once we have designed the costs simulator, we have analyzed routes between Spain and the Black Sea region, considering the imminent entrance of candidate East and Middle European countries into the European Union. Trade operations with all these countries open the door to two big markets: Central Asia and the Middle East. The number of volumes exchanged between Spain and the Black Sea Region shows and upward trend. The total volume of exports and imports between Spain and the Black Sea region is approximately 3,941,806.1 and 24,898,406.1 tons, respectively, value that justifies the management of a trade route between both regions. The data were obtained from figures regarding Spanish import and export operations with Bulgaria, Georgia, Romania, Russia, Turkey and Ukraine, although the countries with the highest number of exchanges with Spain are Greece, Turkey, Russia and Ukraine.

Costs ant times differences for all possible routes between Spain and the Black Sea region have been calculated considering multimodal and road transport. Next tables show the results between different Spanish origins and Black Sea Area.

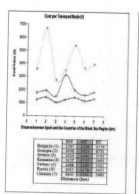
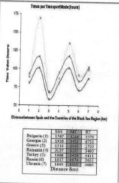

Figure 3: Costs and times differences between multimodal and road transport from Spain to Black Sea region. Source: Own

After the most important exchanges have been selected, Cost Competitiveness Index (CCI) and Time Competitiveness Index (TCI) are calculated. If the resulting value is more than one, then the Short Sea Shipping alternative is more competitive in costs and in time than the only road alternative.

Table 1: Example of Cost Competitiveness Index (CCI) and Time Competitiveness Index (TCI) between Spain and Turkey (Istanbul port). Source: own

	TCI	CCIa	CCIb
Barcelona	1,828896	2,2809114	3,946985
Cádiz	1,643525	1,9559147	2,704304
Madrid	1,746747	2,1564750	3,306404
Murcia	1,848956	2,3061440	3,675953
Valencia	1,937316	2,4842342	4,365663
Vizcaya	1,491680	1,7688517	2,509041
Zaragoza	1,70757	2,0653322	3,175024

Next figure shows the results of Cost Competitiveness Index of SSS versus only road transport between Spain and the Black Sea region routes:

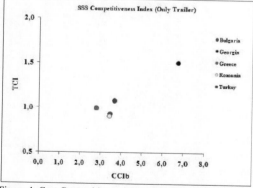

Figure 4: Cost Competitiveness Index of SSS versus only road transport between Barcelona and the Black Sea. Source: Own

From results obtained, we can state:
- The Time Competitiveness Index (TCI) determined that SSS routes between Spain and Georgia and Spain and Ukraine are the most efficient in terms of time.
- The Cost Competitiveness Index (CCIa) determined that SSS routes with the driver, truck and trailer onboard the ship are more competitive in terms of cost than the above case.
- The Cost Competitiveness Index (CCb) determined that SSS routes with only the trailer onboard the ship are the most competitive in terms of cost.

5 CONCLUSIONS AND FURTHER RESEARCH

A design process of a simulator for the assessment of internal and external costs of an only road chain and an intermodal one has been designed. Costs simulator is a fast tool to help customers decide on the most convenient transportation mode for a specific trade link.

This model has been validated according to data from actual commercial exchanges. This validation analysis is quite successful and the simulator data are very close to real prices. The difference between the model data and real data do not exceed 10%.

In this paper we have studied the special case between Spain and the Black Sea Region but the tool presented can assess new maritime trade route comparing with road transport.

A more in deep analysis should be carried out. Data obtained can be used for further research, affording prediction of emissions in the near future by keeping in mind the traffic and fleet evolution and the existing legislation and can also be used to check price variations due to commercial reasons.

REFERENCES

Ametller Malfaz, X. 2007. Optimización del transporte de mercancías mediante el TMCD. UPC Thesis.
Conforti, M. 1992. Productivity scrutinized. 7th Terminal Operations Conference. Cristoforo Colombo Congreso Center. Italy.
De Monie, G. 1989. Medición y evaluación del rendimiento y de la productividad de los puertos. Monographies of the UNCTAD on Port Management, no 6. New York.
Directive 2002 CE, of 18th February 2002,
Dowd, T.J. and Leschine, T.M. 1990. Container terminal productivity: a perspective. Maritime Policy and Management. Vol. 17, no 2, pp.107-112.
European Commission. 2001. The White Paper on Transport: towards 2010. Time to decide. Brussels. http://ec.europa.eu/transport/strategies/2001_white_paper_en.htm
European Commission. 2002/15/EC Directive of the European Parliament and of the council of 22 March on the organiza-

tion of the working times of persons performing mobile activities.

European Commission. Sustainable future for transport: Towards to integrated technology-led and user, COM (2009) 279/4 DGTREN.

Floedstroem, E. 1997. Energy and emission factors for ships in operation. KFB Rep. Swedish Transport and Commerce Res. Board. Swedish Maritime Administration and Mariterm AB. Gothenburg. Sweden.

García Menéndez, et al. 2004. Determinants of mode choice between road and shipping for freight transport. Journal of transport economics and pol- icy. Vol. 38, Part 3.

Martínez de Osés, F.X. and Castells M. 2009. Sustainability of Motorways of the Sea and Fast Ships.

Martínez de Osés, F.X and Castells, M. 2009. Análisis de la aplicación del ecobono en los tráficos marítimos espanoles. Barcelona digital, S.L.

Mulligan, R. Et al. 2006. Short Sea Shipping. Alleviating the environmental impact of economic growth. WMU Journal of Maritime Affaires. Vol. 5, Part 2: 181-194.

Realise Project. Regional Action for Logistical Integration of Shipping across Europe. 2005. AMRIE. 5th Framework Program of the European Union. [http://www.realise-sss.org].

Recordit Project. Real Cost Reduction of door-to- door intermodal Transport. 2001. Supported by the Commission of the European Communities. DGVII. R&DProject, Integrated Transport chains. [www.recordit.org]

Rodríguez Nuevo M et al. 2010. Proposal of a costs simulator, for traffics between Spain and the Black Sea. 3rd International Conference on Maritime and Naval Science and Engineering. Romania

Spanish Ministry of Transport. 2010. Observatorio de costes del transporte por carretera. Secretary of transport, pp. 32-33.

Talley, W.K. 1994. Performance indicators and port performance evaluation. Logistics and Transportation Review, Vol. 30, no4, pp. 339-352.

TERM. Transport and Environment Reporting Mechanism. European Environment Agency. 2002.

Wanhill, Stephen Robert Charles. 1974. Optimum size seaport – a further analysis.

17. Analogical Manoeuvring Simulator with Remote Pilot Control for Port Design and Operation Improvement

P. Alfredini
Escola Politécnica da USP, Sao Paulo, Brazil;Instituto Mauá de Tecnologia, Sao Caetano do Sul, Brazil

J. P. Gerent
Escola Politécnica da USP, Sao Paulo, Brazil

E. Arasaki
Escola Politécnica da USP, Sao Paulo, Brazil; Instituto Nacional de Pesquisas Espaciais, S. J. Campos, Brazil

ABSTRACT: The use of an Analogical Simulator in shiphandling-manoeuvre tests (SIAMA) in waterways constitutes a useful tool for providing improvements in port design and manoeuvring rules, which, when enhanced with other relevant hydraulic studies of Froudian scale models, is a source of valuable statistical information. The time-scale of physical models fast-time runs complie with the square root of the linear scale, in this study-case the model time was 13.04 times faster than prototype. More than 1500 official tests having been undertaken since 1993 by 13 official pilots of three harbours, for manoeuvring and project optimization in 7 piers, with 10 berths, and radio-controlled ore carriers of 75,000, 152,000, 276,000 365,000, 400,000 and 615,000 dwt. The laboratory facilities belong to the Escola Politécnica of Sao Paulo University, Brazil. The port area studied comprised fairways, turning basins and berths. The ships and tugs were unmanned, with tug performance exerted by air fans.

1 INTRODUCTION

On optimizing a harbour lay-out, the relationship between changes in the lay-out, and the resulting changes in flow, wave pattern and the ship's path, should be reproduced as well as possible. In such cases, if small-scale modelling is required, the choice of a physical-model study of the port and adjacent sea area is used. A current or wave-pattern is generated to reproduce local conditions as well as possible. This is the case of SIAMA (Analogical Manoeuvring Simulator) of the Hydraulic Laboratory Harbour and Coastal Division of the São Paulo University, Brazil, which has been fully operational since December 1993. This Brazilian hydraulic centre has clients among governmental and private organisations, maritime authority (Brazilian Navy), port authorities and pilots.

Unfortunately, the risks involved in full-scale measurement, as a research technique, do not permit intense investigation of hazardous manoeuvres in confined channels. Furthermore, full-scale measurement can be undertaken only in existing situations. Excluding field tests, which, besides the high risk involved, are far too expensive, manning models, although requiring large areas, are ideal for training in shiphandling, especially of low-speed control without tug assistance. Real-time simulator studies in a Full Mission Bridge Simulator, although of extreme importance in the design process, only provide limited design-option evaluation, due to

costs and time-consuming if multiple manoeuvre scenarios are to be covered. Consequently only a limited part of pilot experience and knowledge is used for assessing always too few of these options.

The SIAMA analogical concept, with ships and tugs unmanned, is a reliable cost-effective way of combining nautical studies with the hydraulic runs usual in scale-model port-study projects. In fact, this fast-time facility permits running a large amount of test statistics under various desirable environmental conditions (bathymetry, winds, waves, tides and currents), for different vessels and project lay-outs over a limited period.

The aim was to highlight the convincing usefulness of SIAMA, based on more than 1500 official recorded tests in almost 20 years of existence, and different case-studies, thereby encouraging its use in providing information and reducing the number of runs of a Full Mission Bridge Simulator. The ports studied are located in São Marcos Bay (Brazilian north coast): VALE Ponta da Madeira Maritime Terminal (PDM), which is the second in annul loading rate in Latin America (figures 100 Mt) with four piers and seven berths, ALUMAR Harbour (2 berths) and Itaqui Harbour (1 berth) were all studied at this facility.

2 ENVIRONMENTAL CONDITIONS

In Figure 1, it is possible to visualize a tidal current pattern validated through SIAMA. These are difficult conditions for approaching manoeuvres to PDM Pier II, since external currents are greater than those in the confined inner basin. According to Alfredini et al. (2006) and Alfredini et al. (2008), high tidal currents, due to tidal ranges up to 6.5 m, are present in the harbour site studied. The wave-climate is less important, with maximum heights of 1.0 m.

3 DESCRIPTION OF METHODS

The hydraulic model of SIAMA consists of a down-scaled undistorted 1:170 model with 1100 m^2 of the geometry of an estuarine area, in which tidal currents can be generated. In Figure 2A there is an aerial view of VALE PDM Piers I and III (South and North Berths).

Through the simulator, an attempt is made to sail a radio-controlled ship-model along a desired track, with a scenario similar to that in the prototype (see Fig. 2B), reproducing port facilities and navigation aids. Wind effects are provided by fans and tugboat force by air propellers mounted inside the hull of the ship model itself (see Figs 2C an 2D). The ship is steered by pilot-control from ashore (Fig. 2E), using radio-controlled rotating cameras mounted in position on the ship-model bridge (Fig. 2F). Thus, an incorrect visual picture of the waterway can, to a certain extent, be corrected. Since both the vessel and tugs are unmanned, the SIAMA concept is an analogical simulation modeling technique. Furthermore, there is the problem of time: the application of Froude's law to determine scaling factors is required in order to correctly represent total hydraulic influence. This implies that the time-scaling factor is equal to the square root of the linear-scaling factor. Thus, in the model, time will pass faster than in real life (13.04 times in SIAMA). This, depending on the difficulties arising from the quickness of the real manoeuvre, can be a hinderance to the model's pilot. Thus, an apprenticeship for pilot adaptation becomes necessary.

Tests to determine the effect of current forces on the navigation of vessels at port approaches are carried out at SIAMA, by using radio-controlled ships in hydraulic models. Control is exerted over one or two propellers and the ship's rudder. Tug force, when required, are simulated by the thrust of ducted air fans. Six variable speed fans, rigidly fastened inside the hull of the model ship, blow from windows opening thereon (see Fig. 2D). These are located at the bow and stern (longitudinal forces), fore starboard and portside (transversal fore forces), and aft starboard and portside (transversal aft forces). Each fan has an independent speed control, the maximum thrust being 750 kN (prototype). Engine and tugs speeds are calibrated according to prototype data, thereby providing dead slow, slow, half, full ahead, and reverse. In Figure 2G, one can see the propeller and rudder at the stern of a Panamax (75000 dwt) vessel model.

During a test run at SIAMA, a pilot manoeuvres the ship along a variety of courses using any of the available combination of forces. His orders are transmitted by radio communication (Fig. 2E) to a staff controlling tug force, engine speed and rudder angle (see Fig. 2H). Commands are transmitted by radio (see Fig. 2I) to a servo-mechanism on board the vessel model. Simultaneously, photographs are taken from overhead at timed intervals (see Fig. 2J), and video records made by cameras (overhead, vessel-bridge and port). These records show the position and orientation of the ship, as well as the settings of all controls. A later analysis of these records reveals, for example, how tugs need to be used to assist the passage of the ship through the prevailing currents, or whether the forces applied have exceeded those available in practice. Films produced of phenomena occurring in hydraulic models can be shown at reduced speed to enable a true comparison with the full-scale object.

The SIAMA facility has models of ore carriers of 75,000, 152,000, 276,000 365,000, 400,000 and 615,000 dwt. The model-calibration procedure includes turning circle tests compared with prototype data. In Figure 2K, one can see this procedure for a Panamax (75,000 dwt) vessel.

The tests were undertaken by the ports' pilots, with the co-operation and assistance of senior members of the marine departments of the companies concerned. This involved manoeuvres for berthing or departure under various tidal conditions.

In Figure 2, certain features of SIAMA devices are described (for further details see Alfredini et al. 2008). The similarities between bridge manoeuvring vision and prototype conditions are shown. An important issue regarding efficient berthing manoeuvring is berthing force impact, or the equivalent fender deformation, which is measured in gauges (accuracy of 0,01 mm) rigidly mounted into the deck of the pier (see Fig. 2L). Fender stiffness is scaled down by steel blades calibrated according to prototype specifications. The evaluation of impact force in SIAMA runs is used either as an input for fender projects or for pilot training purposes. A graph illustrating deformation in four fenders in a berthing test, is shown in Figure 2M.

Another interesting possibility is to study the conditions of night manoeuvres, since all luminous navigation aids are scaled, so as to comply with the flash time and colors of beacons, warning lights, lighthouses, vessel lights and port lights.

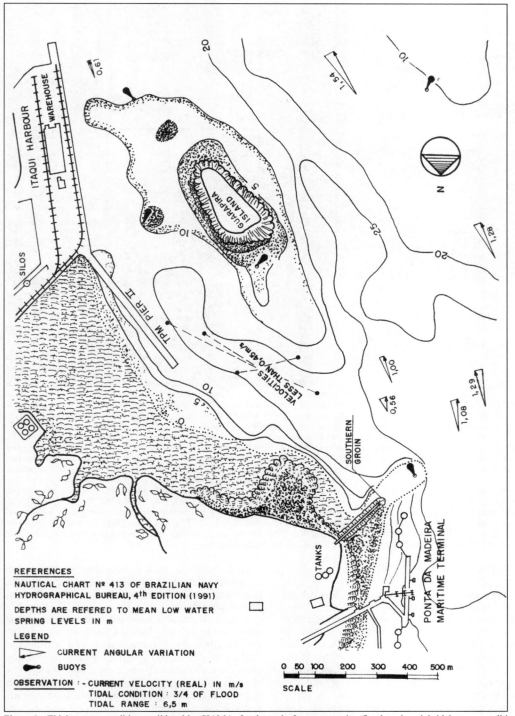

Figure 1. Tidal current conditions validated by SIAMA, for the end of extreme spring flood equinoctial tidal-range conditions, at Ponta da Madeira.

Figure 2. (A) Aerial view of VALE PDM Piers I and III (South and North Berths); (B) SIAMA run of a tug assisted manoeuvre at VALE PDM Pier III. Starboard berthing manoeuvre with a ballast Panamax-class ore carrier (75,000 dwt); (C) Tug assisted manoeuvre at VALE PDM Pier III starboard berthing manoeuvre with a ballast Capesize; (D) Window for an air fan blowing into the Capesize-class ore carrier (152,000 dwt) model hull at fore portside. The calibrated effect reproduces the pushing activity shown in Figure (C); (E) SIAMA run of a tug-assisted approaching manoeuvre for the starboard berthing of a ballast Capesize-class ore carrier (152,000 dwt) at VALE PDM Pier III. One can observe pilot vision from the bridge starboard micro-camera and his radio control device for remote rotating of the camera and for changing the image from the starboard (F) to the portside camera. Orders are also given by radio to tug-masters and engine/rudder (G) SIAMA controlling staff in another room (H); (F) Panamax-class ore carrier (75,000 dwt) vessel model. One can observe the rotating camera mounted in starboard position on the ship-model bridge; (G) Panamax-class ore carrier (75,000 dwt) vessel model stern. One can observe the propeller and rudder; (H) SIAMA controlling staff; (I) The SIAMA radio device; (J) Overhead sequence of photographs of the final stages of a VALE PDM Pier III (North Berth) starboard berthing of a ballast Panamax-class ore carrier (75,000 dwt); (K) SIAMA run turning circle calibration test for a Panamax-class ore carrier (75,000 dwt); (L) View of the gauges mounted on the VALE PDM Pier III North Berth for fender deformation measurement; (M) SIAMA recording of berthing conditions (test showed in J), given by fender deformation in the prototype (in cm), versus manoeuvring prototype time for VALE Pier III (North Berth) fenders;

4 SOME RESULTS AND ANALYSIS

The results of some case studies with SIAMA are hereafter presented with details.

The exceptional conditions with a fairly high level of risk are to be studied: prevailing environmental conditions, the general situation of traffic and failures. Initially, the risk is related to the degree of circumstantial difficulties in the conditions for manoeuvring, e.g. a strong wind, waves and currents. Final risks are related to malfunction in procedures and communications, (misunderstanding, miscalculation and lack of attention) and technical malfunction (engine, rudder, or tug failures). The

consequences, due to the probability of such initial events producing an accident (contact, collision, grounding, etc.) are damage, loss of lives, environmental impacts, etc.

Figure 3 contains a detailed description of the manoeuvring run of a ballast Capesize-class ore carrier (152,000 dwt), for approach and berthing to PDM Pier II, solely with tug assistance, thereby simulating the loss of both engine and rudder. The channel is strongly confined by PDM South Groin (jetty) and the shoals of an island and tidal currents conditions are difficult (see Figure 1). This serves as an example of the usefulness in evaluating the risks of hits or groundings involved in the failure of equipment, such as engine, rudder or tugs, in this case the engine and rudder. As described by Alfredini et al. (2008) and Gerent (2010), for each set of runs the peer group fills in a check-list based upon PIANC et al. 1997, for a de-briefing and open discussion of the following itens:

1 Tug Activity: number of tugs and orders, as well as force employed.
2 Engine Movements: frequency, number.
3 Assessment line and position maintenance: ability to keep the vessel to the intended track, and to assess its position.
4 Position: with relation to the pier and other ships.
5 Control and Safety: feeling of "in control" throughout: feeling of safety.

The following scores were attributed to each item:

10 - practicable, with ease and adequacy
5 - conditionally practicable with certain difficulty
0 – barely practicable

A Pairwise Comparison is obtained, based on this traditional questionnaire. The result of the comparison of every possible pair of alternatives gives a ranking of condition scenarios, viz., much easier, easier, as difficult as, more difficult than and much more difficult than.

From Figure 4, it is possible to gain a quantitative idea regarding the economic impact of a manoeuvring study. In the present case, the results provided an enlargement of daily tidal windows from 6 to 14 hours, with the consensus of a joint group of mariners, engineers and the Brazilian Naval Authority. This study involved all the 13 official pilots, the port captain and naval officers. With the improvements thus obtained, it is now possible to undertake combined manoeuvres (the coordinated approach and berthing of one vessel and the departure of another), by using two additional tugs with more bollard pull (750 kN) than the existing 500 kN. The SIAMA is prepared to make these concomitant manoeuvres with two pilots, with two sets of controls for two vessels. The immediate consequence of this optimization was an increase of 15% in the annual loading rate (equivalent to 10 Mt of iron ore).

Summarizing, SIAMA fast-time simulations comprised the study of more than 400 runs for PDM Pier III, more than 350 for PDM Pier I, more than 300 for the future PDM Pier IV, more than 200 for PDM Pier II, more than 100 for ALUMAR Harbour and more than 50 for Itaqui Harbour. Also important for these statistics was the ideal pilot adaptability to, and familiarisation and cooperation with, SIAMA features. In fact, all the runs were 13.04 times faster, according to the SIAMA Froude law of similitude. As regards the latter, it was also possible to reach a fine balance in the number of runs carried out by each of the official pilots (more than 100 manoeuvres in average). Only under these conditions was it possible to obtain a change in the official manoeuvring rules of the Naval Authorities.

5 CONCLUSIONS

Hydraulic scale-models have an extensive range of application, and it is quite possible that the model of a harbour or river-section is already being built to predict hydrodynamic patterns, pollutant dispersion, mooring conditions, silting, the general arrangement of jetties and breakwaters, etc. In such a case, the choice is whether it is worthwhile to use existing hydraulic models for manoeuvring investigation, that is the SIAMA concept.

An important aspect in design is the lay-out and dimensions of the harbour itself, as well as those of approach channels and turning basins. Harbour efficiency and safety is defined by its nautical accessibility and/or capacity, and hence, economic viability. The strategic dimensioning lay-out and operational-entrance windows in the early design phase can be optimized by fast-time simulations, as those with the SIAMA analogic concept.

In addition to its application as a training tool, SIAMA also facilitates risk analysis in a comparative sense, as a tool for port design. Nevertheless, the main reason for its use is the assistance in simulating the behaviour of those in charge of manoeuvring procedures, normally extremely difficult to simulate in mathematical or other descriptive models. Furthermore, results from fast-time simulator experiments can be incorporated into probabilistic design.

The results presented confirm the possibility of using the SIAMA concept as an cot-effective tool in optimizing port designs, as well as developing the empirical data sets required to sustain further developments in the investigation of manoeuvrability.

ACKNOWLEDGEMENTS

The authors wish to thank the support from USP and Instituto Mauá. Thanks are also due to Vale. Special recognition is mainly due to MSc in Hydraulic Engineering and Master Mariner Captain Joffre Villote, former professor of the Brazilian Ship Simulator, and after PDM Port Captain and Nautical Special Advisor of VALE, for his continuous and encouraging wisdom, and for obtaining the necessary confidence in teamwork, of pilots, masters, tug masters and officers, thus making our task easier.

REFERENCES

Alfredini, P., Amaral, R. F., Araújo, R. N. 2006. Physical model vessel manoeuvring – An analogic Simulator example –

Ponta da Madeira Harbour (Brazil). *First International Conference on the application of physical modelling to Port and Coastal Protection – Coastlab 06*, Oporto, May 2006, IAHR - International Association of Hydraulic Research.

Alfredini, P., Gireli, T. Z., Arasaki, E. 2008. Ships port manoeuvring analogical simulator. *Second International Conference on the application of physical modelling to Port and Coastal Protection – Coastlab 08*, Bari, June 2008, IAHR - International Association of Hydraulic Research.

Briggs, M. J., Melito. I., Demirbilek, Z., Sargent F. 2001. *Deep-draft entrance channels: preliminary comparison between field and laboratory measurements.* Vicksburg: US Army Corps of Engineers ERDC/CHL CHETN-IX-7.

Gerent, J. P. 2010. *A simulação de manobras não tripuladas de navios na otimização de projetos e operações portuárias.* São Paulo: Escola Politécnica of USP MScThesis.

Hensen, H. 1997. *Tug use in port.* London: The Nautical Institute.

PIANC - Permanent International Association of Navigation Congresses 1992. *Capability of ship manoeuvring simulation models for approach channels and fairwaysin harbours.* Brussels: Working Group 20.

PIANC - Permanent International Association of Navigation Congresses, IMPA – International Maritime Pilots Association & IALA - International Association of Lighthouse Authorities 1997. *Approach channels - A guide for design.* Tokyo: Joint PIANC Working Group II-30 in cooperation with IMPA and IALA.

Rowe, R. W. 1996. The shiphandler's guide. London: The Nautical Institut.

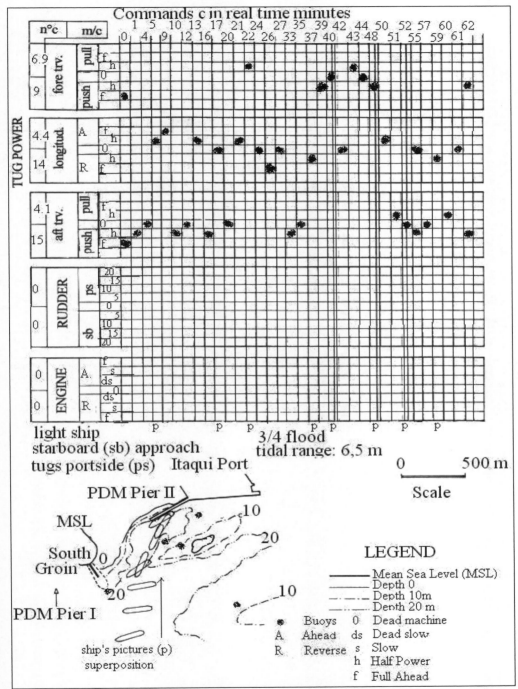

Figure 3. Example of SIAMA run detailed description of a ballast Capesize class ore carrier (152,000 dwt) approaching manoeuvre and berthing at PDM Pier II only with tug assistance, simulating a loss of engine and rudder.

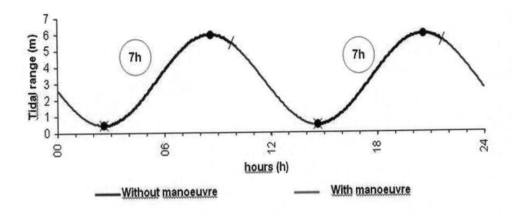

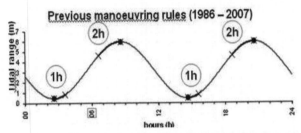

Figure 4. The enlarged tidal windows obtained at SIAMA for PDM Piers I and III. The manoeuvring tidal window was enlarged from 6 to 14 hours daily, thus improving berth simultaneous manoeuvring. The port loading rate increased by 10 million t|year, a 15% improvement

18. A Simulation Model for Detecting Vessel Conflicts Within a Seaport

Q. Li & H. S. L. Fan

School of Civil and Environmental Engineering, Nanyang Technological University, Singapore

ABSTRACT: Conflicts represent near misses between two moving vessels, and often occur in port waters due to limited sea space, high traffic movements, and complicated traffic regulations. Conflicts frequently result in congestion and safety concerns. If conflict risk can be predicted, one could take appropriate measures to resolve conflicts so as to avoid incidents/accidents and reduce potential delays. To the best of this researcher's knowledge, no systematic study has been carried out on the issue of detecting marine traffic conflicts. In this paper, we present an algorithm designed to determine a conflict using the criterion of vessel domain. The algorithm aims to evaluate the relative positions of vessel domains to detect potential conflicts. To implement the algorithm, a simulation model has been developed in Visual C++. The model at present provides a single function for conflict detection but can be expanded to a multi-functional system for resolving conflicts in future work.

1 INTRODUCTION

Traffic conflict refers to the event of vessel interference, which occurs in port waters due to the special characteristics of port traffic in limited sea space, high traffic density, and complex operational regulations. As undesirable incidents, conflicts have a direct effect on the safety of navigation. A conflict without proper resolution may lead to a collision resulting in a loss of life and property, and even threaten the ocean environment.

In recent years marine traffic has been increasing greatly due to the sustained growth of seaborne trade. As a result, the port traffic network becomes finely meshed and intensively used. The demand for the use of sea space sometimes exceeds the available capacity, such that even a small interaction (i.e. a conflict) between vessels may have a large impact on the entire network. The most common product of a conflict is time delay, which results from the evasive maneuvers of vessels to avoid a collision. Within a saturated network, these delays can slow the speed of traffic stream, increase vessel-waiting time and the length of waiting queue. Traffic congestion would arise accordingly.

The world's busiest ports are faced with potential risk of traffic conflicts. However, maritime control centers often can only play an advisory role, which cannot satisfy the demand on traffic management arising within port waters. There is no positive control as to conflict avoidance.

If conflict risk could be predicted in advance, we could take appropriate measures to resolve or eliminate conflicts so as to avoid incidents/accidents and reduce the impact of conflict on network efficiency. However, to the best of this researcher's knowledge, no systematic method has been developed for detecting marine traffic conflicts. A review of past studies related to marine traffic safety revealed that almost all were focused on collision avoidance. Nevertheless, a conflict can be considered as a collision risk with a low degree of danger. Hence, works in collision avoidance are worth reviewing, which could provide valuable reference to this research.

Two criteria are used in past studies for determining a collision risk: the closest point of approach (CPA) and ship domain.

The criterion of CPA is applied with two parameters: distance of closest point of approach (D_{CPA}) and time of closest point of approach (T_{CPA}). The value of CPA parameters indicates the relative position between two vessels. For example, a smaller CPA indicates a higher risk of collision. The CPA parameters are applicable in a collision avoidance system, which can guide vessel to execute proper anti-collision maneuvers. An example is Lenart's studies (Lenart 1999, Lenart 2000) on what speed and/or course maneuver should be undertaken to achieve the required CPA time and distance.

The criterion of CPA is difficult to use in restricted waters such as narrow fairways. In view of this, the concept of ship domain has been proposed as a more comprehensive and accurate criterion. It can be explained as "a water area around a vessel which is needed to ensure the safety of navigation and to avoid collision" (Zhao et al. 1993). Vessel domain was first presented by Fujii et al. (1971). Based on field observations, Fujii's study established a domain model for a narrow channel. Later, Goodwin (1975) developed a domain model in open sea. Besides presenting a model, the study also analyzed how traffic density and length of vessel affect the size of vessel domain.

The shape and size of a vessel domain are affected by a number of factors (vessel's speed and length, sea area, traffic density etc.). As different factors are taken into account, ship domains proposed by various studies differ from one another. Many studies have focused on improving the vessel domain model (Davis et al. 1980, Coldwell 1983, Zhu et al. 2001, Pietrzykowski 2008).

In a port traffic system, vessels traveling along fairways are required to keep various safety clearances in accordance with the port's regulation. The domain of a vessel can thereby be referred to as the clearance area around it. This paper would attempt to design an algorithm to detect conflicts using the criterion of ship domain. That is, the relative positions of the domains of two vessels will be evaluated before they actually encounter each other. If the domain of a vessel will interfere with the domain of the other, a potential conflict is indicated.

A simulation model is developed to implement the algorithm, using Visual C++ 6.0. In the simulation model, conflicts can be detected for a given demand schedule of marine traffic within a seaport. The first and most important goal of conflict detection is to enable safe navigation and avoid collision between vessels. For system optimization, attention should also be paid to reduce the impact of conflicts on network efficiency so as to improve traffic conditions within the seaport.

This paper is structured as follows: Section 1 introduces the issues addressed; Section 2 presents an overview of the simulation model; Section 3 describes the algorithm for conflict detection; Section 4 focuses on simulation model implementation; and Section 5 summarizes findings and proposes future work.

2 OVERVIEW OF SIMULATION MODEL

2.1 The seaport traffic system

A seaport traffic system can be treated as a network of nodes and links. Within the network each link indicates a section of a fairway, and a node can be a berthing/anchorage area, a boarding point for port pilots, an intersection area of fairways, or a separation point dividing a fairway into two sections due to differences in widths and/or traffic regulations. The route of a vessel can be represented by a path in the network consisting of a series of nodes and links.

Figure 1 shows a seaport traffic system we use in the simulation model, where black dots represent the nodes and a rectangle between two nodes indicates a link. The width of a rectangle indicates the width of the link. A vessel is only allowed to travel within the link.

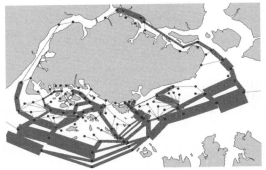

Figure 1. A seaport traffic system for Singapore.

2.2 Flowchart for conflict prediction

A seaport traffic system usually involves a large number of vessels. We need to detect a potential conflict between any pair of vessels. For any pair of vessels, the system will check whether they will conflict or not in a time interval (t_0, t_3).

There are two situations in conflict detection:
– Node conflict prediction: two vessels traveling toward the same node are on different links.
– Link conflict prediction: two vessels traveling toward the same node are on the same link.

In the first situation, the two vessels may have a conflict when they are passing the node. Thus, before the two vessels reach the node, the system needs to predict whether the two vessels will have a conflict.

In the second situation, the two vessels may encounter a conflict on the link. However, if the fairway is sufficiently wide so that a vessel can overtake the other safely, the conflict will not occur. Thus, the factor of the link width should be considered into conflict detection on a link. These are described in the next section.

Note that, the relative position between two vessels varies as vessels are moving. The conflict situation would change accordingly. Suppose that the vessels have a risk of conflict during a certain time period. According to the changes in vessel trajectories, this time period is divided into several time intervals. The system needs to separately evaluate the conflict situation during different time interval.

Figure 2 shows the flowchart for conflict detection (the notations t_0, t_1, t_2, t_3 are defined in Section 3).

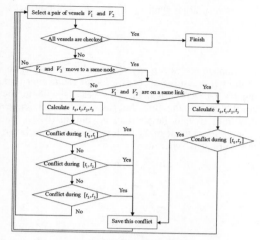

Figure 2. Flowchart for conflict prediction.

3 DETERMINE A CONFLICT BETWEEN TWO VESSELS

3.1 Preliminaries and assumptions

Denote a vessel as V (O, d, Φ, Ψ, $\bar{\Phi}$, $\bar{\Psi}^1$, $\bar{\Psi}^2$) as shown in Figure 3. For the purpose of simplifying analysis, a vessel is regarded as a rectangle V, whose dimensions are Φ (width) and Ψ (length). Suppose $O(x, y)$ is the center of the vessel. At present, it is travelling along the direction d.

The clearance area of a vessel is defined as a zone within which the vessel can keep enough distance to avoid conflict with each other. The clearance area varies according to differences in a vessel's outline, dimension, technical parameters and fairway characteristics. In this research, the shape of a vessel's clearance area is assumed as a rectangle R. The parameter $\bar{\Phi}$ refers to the vessel's lateral clearance. Vessel's longitudinal clearance is represented by parameter $\bar{\Psi}^1$ in the direction of the bow and $\bar{\Psi}^2$ in the direction of the stern. The values of these parameters ($\bar{\Phi}$, $\bar{\Psi}^1$, $\bar{\Psi}^2$) are specified by regulation. These parameters can be set up in a simulation system as input data.

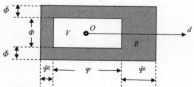

Figure 3. A vessel and its domain.

3.2 Node conflict prediction

Two vessels, V_1 and V_2, on different links travel toward the same node. Table 1 lists the navigation information, where $t_1 < t_2$, i.e. V_1 will reach the node before V_2.

Table 1. Two vessels on different links

	V_1	V_2
Position	A	E
Velocity before the node	v_1	v_2
Velocity after the node	$\bar{v}_1$	$\bar{v}_2$
Time to the node	t_1	t_2
Time to the next node	$\bar{t}_1$	$\bar{t}_2$

Suppose $t_0 = 0$, $t_3 = \min(\bar{t}_1, \bar{t}_2)$. We aim to check whether there is any conflict during the time interval $(0, t_3)$. The movements of V_1 with respect to V_2 are different in three different time intervals

- In the time interval (t_0, t_1), the velocity of V_1 with respect to V_2 is $w_1 = v_1 - v_2$.
- In the time interval $[t_1, t_2]$, the velocity of V_1 with respect to V_2 is $w_2 = \bar{v}_1 - v_2$.
- In the time interval (t_2, t_3), the velocity of V_1 with respect to V_2 is $w_3 = \bar{v}_1 - \bar{v}_2$.

Figure 4 shows the movement of the center of V_1 with respect to the center of V_2. With respect to V_2, starting at A, V_1 passes B at t_1, moves from B to C during $[t_1, t_2]$, and reaches D at t_3. Thus,

$$AB = w_1 t_2 = (v_1 - v_2)t_2,$$
$$BC = w_2(t_1 - t_2) = (\bar{v}_1 - v_2)(t_1 - t_2),$$
$$CD = w_3 t_3 = (\bar{v}_1 - \bar{v}_2)t_3.$$

At location A, the domain of V_1 follows its moving direction v_1 (Fig. 5(a)). Similarly, the domains of the vessels at different locations can be obtained (Table 2). Suppose $q_{ij}^5 = q_{ij}^1$, $i = 0, 1, j = 0, 1, 2, k = 1, 2, 3, 4$. Table 2 tells that

- Q_{ij} is a domain of the vessel V_i at $t = t_i$,
- q_{ij}^k is the k-th corner of the domain Q_{ij},
- $q_{ij}^k q_{k+1}^{ij}$ is the k-th edge of the domain Q_{ij}.

The movement of the domain of V_1 with respect to the domain of V_2 is denoted as the relative movement of V_1 to V_2. For example, referring to Figure 4, Figure 5 shows the relative movements of V_1 to V_2, in the three different time intervals.

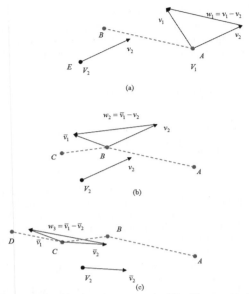

(a)

(b)

(c)

Figure 4. The movements of the center of V_1 with respect to the center of V_2: (a) In time interval $(0, t_1)$; (b) In time interval $[t_1, t_2]$; (c) In time interval (t_2, t_3).

Table 2. Domain of vessels at different locations

Location	Domain of V_1	Domain of V_2
$t=t_0$	$Q_{10}(q_{10}^1,q_{10}^2,q_{10}^3,q_{10}^4)$	$Q_{20}(q_{20}^1,q_{20}^2,q_{20}^3,q_{20}^4)$
$t=t_1$	$Q_{11}(q_{11}^1,q_{11}^2,q_{11}^3,q_{11}^4)$	$Q_{21}(q_{21}^1,q_{21}^2,q_{21}^3,q_{21}^4)$
$t=t_2$	$Q_{12}(q_{12}^1,q_{12}^2,q_{12}^3,q_{12}^4)$	$Q_{22}(q_{22}^1,q_{22}^2,q_{22}^3,q_{22}^4)$

For any j=0, 1, 2, in the time interval (t_j, t_{j+1}), the velocity of V_1 with respect to V_2 is w_{j+1}. The movement of the corner q_{1j}^k with respect to V_2 is a line segment $q_{1j}^k p_{1j}^k$ where

$$p_{1j}^k = q_{1j}^k + (t_{j+1} - t_j)\, w_{j+1}.$$

Thus, the movement of the edge $q_i^k q_i^{k+1}$ with respect to V_2 is $P_j^k = q_1^k q_{1j}^{k+1} p_{1j}^{k+1} p_{1j}^k$ (Fig. 6). If V_1 and V_2 conflict with each other, the movement of at least one edge of V_1 will intersect with the domain of V_2, i.e.

$P_j^k \cap Q_{2j} \neq \emptyset$.

Figure 6 shows an example when there is no conflict between V_1 and V_2. Figure 7 is another example when there is a conflict between V_1 and V_2.

In summary, V_1 and V_2 will conflict in the time interval (t_j, t_{j+1}) if and only if

$\cup(P_j^k \cap Q_{2j}) \neq \emptyset$.

In this way, the conflict detection turns to checking whether two parallelograms intersect with each other or not.

3.3 *Link conflict prediction*

Suppose a vessel V_1 follows another vessel V_2 along a link (see Fig. 8(a)). Table 3 lists the navigation in-

formation of these vessels. The velocity of V_1 with respect to V_2 is

$w_1 = v_1 - v_2$.

If v_1 is not larger than v_2, V_1 and V_2 will conflict if and only if

$$|AE| < \frac{L_1 + L_2}{2}.$$

Suppose $t_3 = \min(t_1, t_2)$. We need to check whether the two vessels will conflict with each other during $(0, t_3)$. After that, the two vessels will not be conflicting on the link, because one vessel leaves this link. If v_1 is larger than v_2, during $(0, t_3)$, the relative movement of V_1 with respect to V_2 is shown in Figure 8(b), where

$$p_1^k = q_1^k + w_1 \times t_3, \qquad k = 1, 2, 3, 4.$$

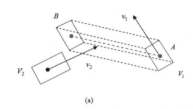

(a)

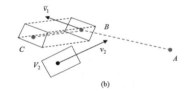

(b)

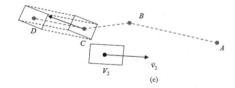

(c)

Figure 5. The relative movement of V_1 to V_2: (a) In time interval $(0, t_1)$; (b) In time interval $[t_1, t_2]$; (c) In time interval (t_2, t_3).

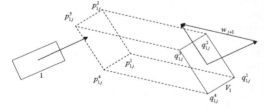

Figure 6. $P_j^k \cap Q_{2j} \neq \emptyset$, V_1 and V_2 will not conflict with each other in the time interval (t_j, t_{j+1}).

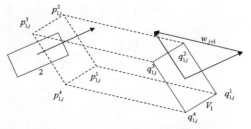

Figure 7. $P_j^2 \cap Q_{2j} \neq \emptyset$, $P_j^3 \cap Q_{2j} \neq \emptyset$, V_1 and V_2 will conflict with each other in the time interval (t_j, t_{j+1}).

Obviously, V_1 and V_2 will have a conflict if and only if $q_1^1 q_1^4 p_1^2 p_1^3$ intersects with Q_2. In Figure 9(a), $q_1^1 q_1^4 p_1^2 p_1^3$ intersects with Q_2, thus V_1 and V_2 are in conflict. In Figure 9(b), $q_1^1 q_1^4 p_1^2 p_1^3$ does not intersect with Q_2, thus V_1 and V_2 will not conflict with each other.

The conflict detection method described earlier is merely based on an assessment of relative movement of one vessel to another vessel. The system judges if a conflict will occur by checking whether the two parallelograms intersect. If link width is taken into consideration, the result may be different. Suppose the width of the link is W, and the width of the domains of the two vessels are W_1 and W_2. The two vessels can travel in parallel without a conflict if the width of the link is not smaller than the sum of W_1 and W_2 (Fig. 10).

Table 3. Two vessels on different links

	V_1	V_2
Position	A	E
Velocity	v_1	v_2
Domain	$Q_1 (q_1^1, q_1^2, q_1^3, q_1^4)$	$Q_2 (q_2^1, q_2^2, q_2^3, q_2^4)$
Domain length	L_1	L_2
Domain width	W_1	W_2
Time to leave	t_1	t_2

(a)   (b)

Figure 8. Vessels on the same link.

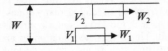

(a)   (b)

Figure 9. Vessel conflict on the same link depends on initial vessel positions

Figure 10. Vessels travel in parallel on a link

4 EXAMPLES AND DISCUSSIONS

We have implemented the conflict detection algorithm in the simulation model using Visual C++ 6.0 on a Windows XP operating system. In this section, some examples are shown to illustrate the results of our algorithm.

Figure 11 is the first example. Two vessels travel toward the same node. A vessel is represented as a rectangle with a solid line indicating the travelling direction. The vessel on the left hand side is V_1 and the other one is V_2. The gray areas in Figure 11(a) indicate the link areas. At any time, a vessel keeps inside a link area. The gray areas in Figure 11(b), enclosed by solid lines, indicate the relative movement of V_1 to V_2. A conflict is predicted since the relative movement intersects with the domain of V_2 (Fig. 11(b)). Figure 11(e) shows that the two vessels conflict with each other when they are passing the node.

The result also shows that the conflict can be predicted at any time before the conflict time. Figure 11(b) and Figure 11(d) show the relative movements of V_1 to V_2 at different positions. As we can see from the figure, the same conflict is predicted in both positions. In fact, the conflict prediction algorithm will detect the conflict any time before either vessel reaches the node. The result implies that we can increase the simulation time interval, thus reduce the calculation for conflict detection.

Figure 12 is an example with two vessels travelling on the same link. The vessel V_1 tries to catch up with V_2. The gray area enclosed by black lines in Figure 12(b) is the relative movement of V_1 to V_2. Combining Figure 12(a) and Figure 12(b), we know that this relative movement intersects with the domain of V_2. Thus V_1 will catch up with V_2. If the link with is not enough, V_1 and V_2 will come into conflict (Fig. 12(c)). On the other hand, if the link width is enough for the two vessels to navigate in parallel, there will be no conflict (Fig. 13).

In Figure 13(a), two vessels in parallel will not conflict with each other, because the relative movement of V_1 to V_2 does not intersect with the domain of V_2. Therefore, V_1 can catch up with V_2 and overtake it.

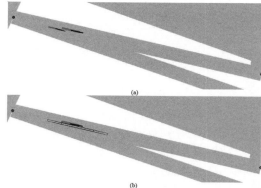

Figure 13. Two vessels on the same link travel in parallel.

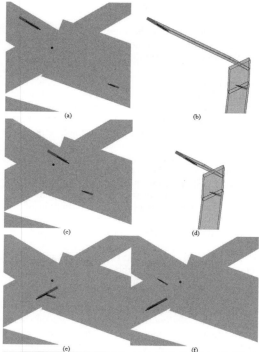

Figure 11. Two vessels from different links conflicting each other at a node

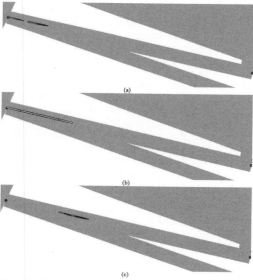

Figure 12. Two vessels on the same link in conflict

5 SUMMARY AND CONCLUSIONS

A simulation model has been developed for predicting potential vessel conflicts within a seaport. An algorithm for conflict detection was designed with the use of ship domain criterion: when the relative movement of one vessel with respect to a second vessel intersects with the domain of the second vessel, the two vessels will have a conflict. The algorithm simplifies the conflict detection problem by checking whether two parallelograms intersect with each other.

An application of the model was demonstrated using the seaport of Singapore as an example. Inputs to the model include the background map, data on fairways, and information on vessel types and characteristics. Vessel arrivals and vessel routes are generated by the model according to statistical distributions. Simulation results showed that conflicts can be accurately predicted in time. The logic of conflict detection is applicable to other traffic systems by changing the input data. Thus, the simulation model is a generic model which can be adapted to other busy seaports that are faced with traffic congestion and delays.

For future work, human factor could be taken into account. Human error would affect vessel movement as well as conflict situation. An example is the situation where one vessel follows another vessel along a link. Even with sufficient width for overtaking, an accident/ incident may occur as a result of human error. Another possible improvement will focus on the determination of a reasonable time step in simulation. In the current system, a single conflict may be detected for multiple times. When the time step is too small, a conflict may be predicted many times. On the other hand, if it is too large, some conflicts may not be detected. Thus, the determination of an optimum simulation time step is also an important issue.

REFERENCES

Coldwell, T.G. 1983. Marine Traffic Behaviour in Restricted Waters. *Journal of Navigation* 36(3): 430-444.

Davis, P.V., et al. 1980. A Computer Simulation of Marine Traffic Using Domains and Arenas. *Journal of Navigation* 33(2): 215-222.

Fujii, Y. and Tanaka K. 1971. Traffic Capacity. *Journal of Navigation* 24(4): 543-552.

Goodwin, E.M. 1971. A statistical study of ship domains. *Journal of Navigation* 28: 328-344.

Lenart, A.S. 1999. Manoeuvring to required approach parameters - CPA distance and time. *Annual of Navigation* 1: 99–108.

Lenart, A.S. 2000. Manoeuvring to Required Approach Parameters - Distance and Time Abeam. *Annual of Navigation* 2: 81–88.

Pietrzykowski, Z. 2008. Ship's Fuzzy Domain-a Criterion for Navigational Safety in Narrow Fairways. *Journal of Navigation* 61(03): 499-514.

Zhao, J., et al. 1993. Comments of Ship Domains. *Journal of Navigation* 46: 422-436.

Zhu, X.L., et al. 2001. Domain and Its Model Based on Neural Networks. *Journal of Navigation* 54(1): 97-103.

19. Research on Ship Navigation in Numerical Simulation of Weather and Ocean in a Bay

T. Soda
Kobe University, Faculty of Maritime Sciences

S. Shiotani
Kobe University, Organization of Advanced Science and Technology

H. Makino & Y. Shimada
Kobe University, Faculty of Maritime Sciences

ABSTRACT: For safe navigation, high-resolution information on tidal current, wind and waves is very important. In coastal areas in particular, the weather and ocean situation change dramatically in time and place according to the effects of geography and water depth. In this paper, high-resolution wave data was generated using SWAN as a numerical wave model. To estimate waves, wind data is necessary. By using the mesoscale meteorological model of WRF-ARW, detailed wind data was generated. The tidal current data was generated by using POM.

We simulated tidal currents, wind and ocean waves for the duration of a typhoon passing over Japan in September of 2004.

Secondly, we simulated ship maneuvering using simulated tidal current, wind and wave data. For the ship maneuvering model, the MMG (Mathematic Modeling Group) was used. Combining high-resolution tidal current, wind and wave data with the numerical navigation model, we studied the effects of tidal current, wind and waves on a ship's maneuvering. Comparing the simulated route lines of a ship with the set course, it was recognized that the effects of the tidal current, wind and waves on a moving ship were significant.

1 INTRODUCTION

Winds, tidal currents, and waves of the ocean are considered the most important factors in the field of ocean engineering. In particular, the numerical forecasting of tidal currents, winds over the sea, and ocean waves in coastal areas is important for ocean environments, fisheries, and navigation. In the previous paper[1][2], the numerical simulation of tidal currents as they relate to the winds in a bay was explained. The simulation of tidal currents was carried out in Japan's Osaka Bay. Detailed tidal currents were calculated using the POM (Princeton Oceanography Model). The results of the numerical simulation of tidal currents and tidal elevations were compared with data from observations in the bay. A comparison of the numerical and calculated results showed agreement.

In this paper, a numerical simulation of winds on the sea was carried out in Japan's Osaka Bay area. Details of the distribution of winds on the sea were calculated using the WRF model developed principally by National Centers for Environmental Prediction (NCEP) and National Oceanic and Atmospheric Administration (NOAA) in the United States. The wind calculation was continued for 4 days while a typhoon passed over the Nihon Sea near Japan. A strong wind blew from the south on the coastal sea area. The simulation of ocean waves was carried out in the same bay area where the wind simulation was done, and the calculated wind and tidal currents were used. Details of the distribution of waves on the sea were calculated using the SWAN[3][4] (Simulating WAves Nearshore) model developed at Delft University of Technology in the Netherlands. The simulation of tidal currents was calculated using the POM numerical model. We analyzed wind stress on tidal current by using WRF-calculated wind data.

Secondly, the numerical simulation of winds and waves was applied to a navigational simulation of a sailing ship in the bay area.

The accurate estimation of a given ship's position is very important for optimal ship routing[5]. Such estimations can be obtained when the hydrodynamic model, which is widely used to describe a ship's maneuvering motion, is adopted in order to estimate a ship's position. As a first step toward this final objective of optimum routing, the effects of winds, waves, and tidal currents on a ship's maneuverability were examined through numerical simulations.

2 NUMERICAL SIMULATION OF OCEAN WINDS

The simulation of winds was carried out by WRF-ARW3.1.1, a mesoscale meteorological model developed principally among the National Center for Atmospheric Research (NCAR), the National Oceanic and Atmospheric Administration (NOAA), the National Centers for Environmental Prediction (NCEP), the Forecast Systems Laboratory (FSL), the Air Force Weather Agency (AFWA), the Naval Research Laboratory, the University of Oklahoma, and the Federal Aviation Administration (FAA).

The equation set for WRF-ARW is fully compressible, Eulerian, and nonhydrostatic, with a run-time hydrostatic option. The time integration scheme in the model uses the third-order Runge-Kutta scheme, and the spatial discretization employs second- to sixth-order schemes.

GFS-FNL data were used as boundary data[6]. The Global Forecast System (GFS) is operationally run four times a day in near-real time at NCEP. GFS-FNL (Final) Operational Global Analysis data are set on 1.0 x 1.0 degree grids every 6 hours.

The simulated term was 96 hours from 5 September, 2004, 00:00 UTC to 9 September, 2004, 00:00 UTC. Figure 1 shows the weather charts of the simulated term. In this figure, (b) shows the typhoon located over the southwest of Japan on 7 September 00:00 UTC, and (c) shows the area after the typhoon had passed on 8 September 00:00 UTC.

Two areas for nesting were calculated in order to simulate winds accurately. While the typhoon was passing over Japan, a strong south wind blew on the Japanese Pacific side. Figure 2 shows the two areas, d01 and d02. The center point of d01 is E135.52 N34.72.

The numerical simulation of wind was carried out in the area around Japan. The grid numbers are 44 x 35 x 28 in the x-y-z axis in d01 and 41 x 36 x 28 in the x-y-z axis in d02. The horizon grid intervals of Δx and Δy are 10 km in d01 and 2 km in d02. In both areas, the vertical grid is 20 from top pressure (500 Pa) to ground pressure. The condition calculated by WRF is shown in Table 1.

At the points shown in Figure 3, the calculated wind data were verified by the wind observed with the AmeDAS, the system of the Japan Meteorological Agency. The results of wind simulation at these points are shown in Figure 4. The horizontal axis shows the time from the start time of calculation in hours. The vertical axis shows the wind velocity and wind direction.

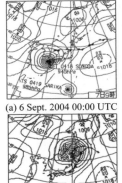

(a) 6 Sept. 2004 00:00 UTC

(b) 7 Sept. 2004 00:00 UTC

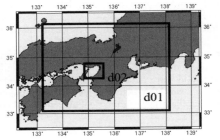

(c) 8 Sept. 2004 00:00UTC

Figure 1. Chart of calculation term

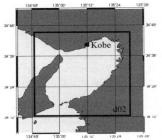

Figure 2. Two calculation areas

Table 1. Condition of calculations by WRF

	d01	d02
Dimension	44 x 35 x 28	41 x 36 x 28
Mesh size	10 (km)	2 (km)
Time step	60 (s)	12 (s)
Start time	2004-09-05-00:00:00 UTC	
End time	2004-09-09-00:00:00 UTC	

Figure 3. The comparison points for wind, tide level and wave height

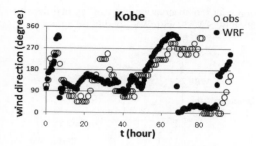

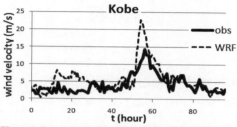

Figure 4. Comparison of calculated and observed wind velocities and directions

The results of the wind simulation are shown in Figure 4. In the time when the typhoon was closest, the estimated wind velocity is lower than the observed data at the point in Kobe. The RMS error of wind velocity is 2.3 m/s, and the correlation coefficient of wind velocity is 0.89. The simulation of wind is generally estimated accurately.

3 NUMERICAL SIMULATION OF TIDAL CURRENT

The estimation of tidal current was carried out by using the Princeton Oceanographic Model (POM) (Mellor 2004). The basic equations of the tidal current are the continuity equation and Navier-Stokes equation, shown as follows:

$$\frac{\partial DU}{\partial x} + \frac{\partial DV}{\partial y} + \frac{\partial \omega}{\partial \sigma} + \frac{\partial \eta}{\partial t} = 0 \qquad (1)$$

$$\frac{\partial UD}{\partial t} + \frac{\partial U^2 D}{\partial x} + \frac{\partial UVD}{\partial y} + \frac{\partial U\omega}{\partial \sigma} - fVD + gD\frac{\partial \eta}{\partial x}$$
$$+ \frac{gD^2}{\rho_0}\int_\sigma^0 \left[\frac{\partial \rho'}{\partial x} - \frac{\sigma'}{D}\frac{\partial D}{\partial x}\frac{\partial \rho'}{\partial \sigma'} \right] d\sigma' = \frac{\partial}{\partial \sigma}\left[\frac{K_M}{D}\frac{\partial U}{\partial \sigma} \right] + F_x \qquad (2)$$

$$\frac{\partial UD}{\partial t} + \frac{\partial UVD}{\partial x} + \frac{\partial V^2 D}{\partial y} + \frac{\partial V\omega}{\partial \sigma} + fUD + gD\frac{\partial \eta}{\partial y} +$$
$$\frac{gD^2}{\rho_0}\int_\sigma^0 \left[\frac{\partial \rho'}{\partial y} - \frac{\sigma'}{D}\frac{\partial D}{\partial y}\frac{\partial \rho'}{\partial \sigma'} \right] d\sigma' = \frac{\partial}{\partial \sigma}\left[\frac{K_M}{D}\frac{\partial V}{\partial \sigma} \right] + F_y \qquad (3)$$

where (u and v) are the components of the horizontal velocity of tidal current, ω is the velocity component of the normal direction to the σ plain, f is the Coriolis coefficient, g is the acceleration of gravity, F_x and F_y are the horizontal viscosity diffusion coefficients, and KM is the frictional coefficient of the sea bottom.

We calculated the effect of wind stress on tidal current by using wind data gathered by WRF. The grid number is 328 x 288 in the x-y axis in d02. The horizon grid interval of Δx and Δy is 250 m in d02. The calculation time interval was 2 seconds.

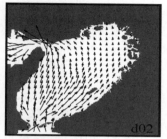

(a) Flow 2004-09-07 04:00 UTC

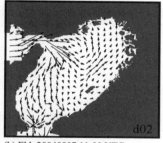

(b) Ebb 20040907 11:00 UTC

Figure 5. Distribution of calculated tidal current on the surface of the sea

Figure 5. Surface tidal current when the typhoon was closest. A comparison of flow and ebb shows that the direction of the tidal current was changed dramatically.

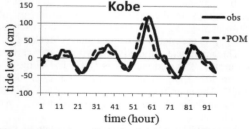

Figure 6. Comparison of calculated and observed tidal level

The tidal level of the simulated and observation in Kobe are shown in Figure 4. The tidal level during the typhoon's approach was higher by strong southern wind. The tidal level was estimated accuracy.

4 NUMERICAL SIMULATION OF OCEAN WAVES

As a numerical model for simulating waves, we used SWAN, a third-generation wave simulation model developed at Delft University of Technology.

The SWAN model is used to solve spectral action balance equations without any prior restriction on the spectrum for the effects of spatial propagation, refraction, reflection, shoaling, generation, dissipation, and nonlinear wave-wave interactions. For the SWAN model, the code used was the same as that used for the WAM model. The WAM model calculates problems in deep water on an oceanic scale, and SWAN considers problems from deep water to the surf zone. Consequently, the SWAN model is suitable for estimating waves in bays as well as in coastal regions with shallow water and ambient currents.

Information about the sea surface is contained in the wave variance spectrum or energy density $E(\sigma,\theta)$. Wave energy is distributed over frequencies σ and propagation directions θ. σ is observed in a frame of reference moving with the current velocity, and θ is the direction normal to the wave crest of each spectral component. The action balance equation of the SWAN model in Cartesian coordinates is as follows:

$$\frac{\partial N}{\partial t}+\frac{\partial}{\partial x}(C_x N)+\frac{\partial}{\partial y}(C_y N)+\frac{\partial}{\partial \sigma}(C_\sigma N)+\frac{\partial}{\partial \theta}(C_\theta N)$$
$$=\frac{S}{\sigma} \qquad (4)$$

where the right-hand side contains S, which is the source/sink term that represents all physical processes which generate, dissipate, or redistribute wave energy. The equation of S is as follows:

$$S = S_{in} + S_{ds,w} + S_{ds,b} + S_{ds,br} + S_{nl4} + S_{nl3} \qquad (5)$$

in the right-hand side, where S_{in} is the transfer of wind energy to the waves, $S_{ds,w}$ is the energy of whitecapping, $S_{ds,b}$ is the energy of bottom friction, and $S_{ds,br}$ is the energy of depth-induced breaking.

The numerical simulation of waves was carried out in d02. The grid number is 164 x 144 in the x-y axis in d02. The grid interval of Δx and Δy is 500 m in d02. The conditions for calculation by SWAN are shown in Table 2. Figure 7 is a flowchart of wave calculations.

The results of wave simulation, shown in Figure 8, include the significant wave height developing during the typhoon's approach. Comparing the simu-

lation with the observation, we can say that the simulation by SWAN with WRF and POM agrees.

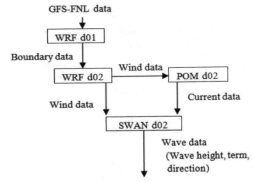

Figure 7. Flow chart for wave calculation by SWAN

Table 2. Conditions for calculations by SWAN

	d02
Dimension	164 x 144
Mesh size	500 (m)
Time step	15 (min)
Start time	2004-09-05-00:00:00 UTC
End time	2004-09-09-00:00:00 UTC
Number of frequencies	30 (0.04Hz-1Hz)
Number of meshes in θ	36

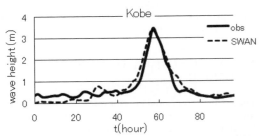

Figure 8. Comparison of calculated and observed wave heights

5 SHIP MANEUVERING SIMULATION

The accurate estimation of a ship's position is very important for optimum ship routing. Such estimations can be obtained when hydrodynamic forces and moments affecting the hull are known in advance. The MMG (Mathematical Model Group) model, widely used to describe a ship's maneuvering motion, was adopted for the estimation of a ship's location by simulation[7]. The primary feature of the MMG model is the division of all hydrodynamic forces and moments working on the vessel's hull, rudder, propeller, and other categories, as well as the analysis of their interaction. The coordinate system is denoted in Figure 9.

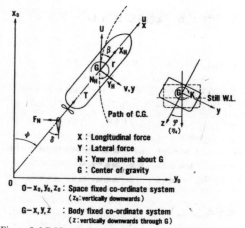

X : Longitudinal force
Y : Lateral force
N : Yaw moment about G
G : Center of gravity

$O-x_0, y_0, z_0$: Space fixed co-ordinate system
(z_0: vertically downwards)

$G-x, y, z$: Body fixed co-ordinate system
(z : vertically downwards through G)

Figure 9. MMG coordinate system

Two coordinate systems, space-fixed and body-fixed, are used in ship maneuverability research. The latter system, G-x,y,z, moves together with the ship and is used in the MMG model. In this coordinate system, G is the center of gravity of the ship, the x-axis is in the direction of the ship's course, the y-axis is perpendicular to the x-axis on the right-hand side, and the z-axis runs downward vertically through G.

Therefore, the equation for the ship's motion in the body-fixed coordinate system adopted in the MMG model is written as follows:

$$(m+m_x)\dot{u} - (m+m_y)vr = X$$
$$(m+m_y)\dot{v} + (m+m_x)ur = Y \qquad (6)$$
$$(I_{zz} + J_{zz})\dot{r} = N$$

where m is the mass, the m_X and m_Y areas are the added mass, and u and v are the components of the velocity in the direction of the x-axis and y-axis, respectively. r is the angular acceleration. I_{ZZ} and J_{ZZ} are the moment of inertia and the added moment of inertia around G, respectively. X and Y are the hydrodynamic forces, and N is the moment around the z-axis.

According to the MMG model, the hydrodynamic force and the moment in the above equation can be shown as follows:

$$X = X_H + X_P + X_R + X_A + X_W + X_E$$
$$Y = Y_H + Y_P + Y_R + Y_A + Y_W + Y_E \qquad (7)$$
$$N = N_H + N_P + N_R + N_A + N_W + N_E$$

where the subscripts H, P, R, A, W and E denote the hydrodynamic force or moment induced by the hull, propellar, rudder, air, waves and external forces, respectively.

The hydrodynamic forces caused by wind are defined in Equation (8):

$$X_A = \frac{\rho_A}{2} V_A^2 A_T C_{XA}(\theta_A)$$
$$Y_A = \frac{\rho_A}{2} V_A^2 A_L C_{YA}(\theta_A) \qquad (8)$$
$$N_A = \frac{\rho_A}{2} V_A^2 L A_L C_{NA}(\theta_A)$$

where ρ_A is the density of air, θ_A is the relative wind direction, V_A is the relative wind velocity, and A_T and A_L are the frontal projected area and the lateral projected area, respectively. C_{XA}, C_{YA}, and C_{NA} are the coefficients, which were estimated by the method of Fujiwara et al.[8]

The hydrodynamic forces caused by waves are defined as follows:

$$X_W = \rho g h^2 B^2 / L \overline{C_{XW}}(U, T_V, \chi - \psi_0)$$
$$Y_W = \rho g h^2 B^2 / L \overline{C_{YW}}(\omega_0, \chi - \psi_0) \qquad (9)$$
$$N_W = \rho g h^2 B^2 / L \overline{C_{NW}}(\omega_0, \chi - \psi_0)$$

where ρ is the density of seawater, g is the acceleration of gravity, h is the amplitude of significant wave height, B is the ship's breadth, L is the length of the ship, and $\overline{C_{XW}}$, $\overline{C_{YW}}$, and $\overline{C_{NW}}$ are averages of short-term estimated coefficients. The hydrodynamic force on the hull surface, including the added resistance, wave-induced steady lateral force, and yaw moment, was obtained through the Research Institute on Oceangoing Ships (RIOS) at the Institute of Naval Architecture, Osaka University. The RIOS was established for the purpose of improving the performance of ships in wind and waves[9].

The frequency-domain response characteristics of wave-induced ship motions with six degrees of freedom were computed using the principal proporties, arrangement plan, and body plan of the ship. In the RIOS system, the wind wave is represented by the ITTC spectrum, and the swell is represented by the JONSWAP spectrum. In this study, the average added resistances, wave-induced steady lateral forces, and yaw moments to the ship by wind-wave and swell are combined.

We simulated the maneuvering of the training ship *Fukaemaru*, the main characteristics of which are shown in Table 3. The relevant data describing the manoevering is shown in reference 2. The starting point of the maneuvering simulation was N34.4 E150, and the set course was 50 degrees. The term of simulation was 4800 seconds from 2004-09-08 05:00:00 UTC. Figure 10 shows the distribution of the tidal current at 2004-09-08 05:00:00 UTC and set ship course. The numerical navigation was carried out at a fixed propeller revolution of 9.0 kn in still water (revolution is 500 rpm).

Table 3. Main characteristics of Fukaemaru

Loa	49.95 m
Lpp	45.00 m
Breadth	10.00 m
Depth	6.10 m
Draft	3.20 m
Gross Tonnage	449 ton
Main Engine Output	1,100 kw
Trial Speed	14.28 knots
Sea Speed	12.50 knots
Steering Engine	3.7 kw

Figure 10. Distribution of tidal current at 2004-09-08 05:00 UTC and ship course

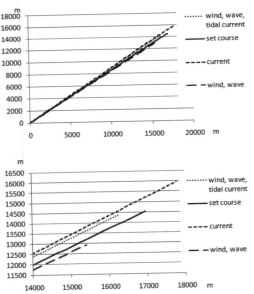

Figure 11. Comparison of wind, wave, and tidal current effect

Figure 11 is the simulated ship course from the start point. In the illustrations below Figure 11, the lower illustration magnifies the rectangular area of the upper one. Each line shows the track based on the effects of "wind and wave," "tidal current," "wind, wave, and tidal current," and the route setting. The results of the numerical navigation, including the effects of winds, waves, and tidal currents, were examined.

6 CONCLUSION

In the present basic study of a numerical navigation system for an oceangoing ship in a bay area, the effects of winds, waves, and tidal currents were studied. The main conclusions are as follows:
1 By combining the numerical models of WRF, POM and SWAN, accurate, high-resolution tidal current, wind and wave data were generated.
2 By using the detailed data, wind and wave force upon the ship was estimated.
3 Detailed data on winds, waves, and tidal currents were applied to a numerical navigation model with estimations of force upon a sailing ship.
4 The effects of winds, waves, and tidal currents on a ship's maneuverability were significant.

With the above information, it is possible to achieve an optimum route by utilizing a numerical simulation if winds, waves, and tidal currents can be predicted in a bay area.

REFERENCES

[1] Xia H., Shiotani S., et al. (2005). A Study of Weather Routing Considering Real Time Data of Weather and Ocean for Sailing Ship in Coastal Area – Basic Simulation of Ship Positioning by Ship Maneuvering and Experiment by a Real Ship, Journal of the Kansai Society of Naval Architects, Japan, No. 234, pp. 159-166.
[2] Xia H., Shiotani S., et al. (2006). Estimation of Ship's Course for Sailing on Route by Navigation Simulation in Coastal Water, The Journal of Japan Institute of Navigation, No. 115, pp. 51-57.
[3] Booji N., et al. (1999). A Third-generation Wave Model for Coastal Regions, Part I , Model Description and Validation, J. Geograph. Research, 104, C4, 7649-7666.
[4] The SWAN Team (2009). SWAN, Scientific and Technical Documentation, SWAN Cycle III version 40.72ABC, http:/www.swan.tudelft.nl, Delft University of Technology.
[5] Shimada Y., Shiotani S., Takahashi K. (2010). Influence of Ocean Model Horizontal Resolution on Weather Routing, Techno-ocean, CD
[6] Mase H., Katsui S., Yasuda T., Tomand T. H., Ogawa K. (2006). Verification of GFS-WRF-SWAN Wave Prediction System by Three Seasons' Comparison, Annual Journal of Civil Engineering in the Ocean, JSCE VOL. 22, pp. 109-114.
[7] Yoshimura Y. (1986). Mathematical Model for the Maneuvering Ship Motion in Shallow Water, Journal of the Kansai Society of Naval Architects, Japan, No. 200, pp. 41-51.
[8] Fujiwara T., Ueno M., Nimura T. (1998). Estimation of Wind Forces and Moments Acting on Ships, Journal of the Society of Naval Architects of Japan (183), 1998-06, pp. 77-90.
[9] The Research Initiative on Oceangoing Ships (2011). http://133.1.7.5/

Navigational Systems and Simulators – Marine Navigation and Safety of Sea Transportation – Weintrit (ed.)

Navigational Simulators

20. A Methodological Framework for Evaluating Maritime Simulation

P. Vasilakis & N. Nikitakos

University of Aegean Department of Shipping Trade & Transportation, Chios, Greece

ABSTRACT: The application of simulation courses according to STCW conversion is addressed to the education of marine deck and engine officers in order to familiarize with the working environment, emergency contingency training and trouble shooting. This paper presents a framework which evaluates the participants in the courses of simulator, according to their concerns and their level of use. Actually this framework is an innovative concept which tries to identify how the contributors think and work in this virtual environment. The results from the application of this framework are presented in this paper, based on student's concerns, reactions and level of use with respect to the exercise and efficiency of simulation training courses taken place at the Merchant Marine Academy of Engine Officers on Chios Island. The main goal of our research is to promote a general framework which can be easily applied in any marine simulation curses and will be very useful to the instructor for reorganizing, redesigning and finally configurating the Simulation Courses according to their participant.

1 INTRODUCTION

According to STCW section A-I/12"..... the simulator shall be capable of simulating the operating capabilities of shipboard equipment , to a level of physical realism appropriate to the training and assessment objectives.....".

In trying to identify the capability of maritime simulator course, we can say that is multifunction's tool through which special techniques in ships handling either to deck or to engine department can be promote. More specifically, we can say that maritime simulator can work a knowledge accelerator for the seafarer in order to protect the human life at sea and the environment protection.

Actually the simulator course is an interactive course among the machinery, the instructor and the students. But how do those three ingredients interact with each other, under which pedagogic model? And finally who makes the evaluation and the assessment of this course?

Those are some of the main questions which we will try to answer through our research.

The main goal in most of the pedagogic theories is what finally the students take from the course in long term future and not by the end of the course. From literature view the closer pedagogic theories to the virtual learning like maritime simulation are: Problem based learning, Discovering Learning, Learning by exploration, just in time teaching, Case based learning. The main guide line in all this theory is that the evaluation of the course stems from the interaction between the student and the instructor.

Based on this acceptance, in our research we will try to apply the Concern Based Model in Full Mission Engine Room Simulator which has steadily been introduced and corresponded to a fraction of the academic syllabus in the merchant marine academy of Chios. This Study is a part of an expanded research according to the Concerned Based Adoption Model (CBAM) which includes the:

– Stage of Concerned
– Level of Use

Within this paper, the value of simulation along the stage of concern and the Level of Use of simulator taught courses will be examined. Data are extracted from student reactions that have participated and experienced simulator applications. The analysis was based on the theoretical model C.B.A.M. Results are extracted and anticipated to present the perception skills of participating students as well as their reactions and feedback.

The scope of the analysis is to obtain results that will enable the academic staff to ascertain the needs and requirements for improvements in order to develop more resourceful and efficient methods for simulation course education. Thus the application of simulators will be far more productive for the participating students enabling them also to be adapted in the real working environment.

The objective goal of application of simulator courses is the reproduction of virtual reality cases/problems that converge to the real operation techniques and troubleshooting that the officers might encounter during real operation at the ships engine room.

By evaluating the performance of simulator courses it is anticipated and expected to improve the offered education; thus the candidate officers will make a success when they face the "real world". It will be a virtue and success of the academic schools that are associated with the maritime education to graduate qualified officers who are not only qualified graduates but they also bare the knowledge and ability to undertake and effectively execute their duties without any doubt and with the expected professionalism and responsibility.

2 PARTICIPANTS

The participants who took part in the CBAM questionnaire are students of Merchant Marine Academy on Chios Island. The Academy was founded in 1955 and currently the supported by the Greek Ministry of Economy, Competitiveness & Shipping. The duration of studies is 4 years amongst which 12 months are practice on merchant marine vessels. The rest three years consist of a syllabus of theoretical and laboratory courses. The Orientation of Students is both theoretical and technical education according to Sandwich Courses and STCW. Graduates are granted with the diploma of 3rd Engineer of Merchant Marine.

By total of sixty five students that participated in the course of simulation during spring (half-year) period 2008-2009, fifty questionnaires were collected in total. Nine questionnaires were not fully completed thus they were excluded from the analysis.

In order to present a complete picture for the standards and knowledge of the participating students the following are noted:

– The students are in a percentage of 60% of 20-22 years if age.
– Most students came from high school with general education, in a percentage of 46, 3%. This fact determines that they are individuals with high level of theoretical knowledge.
– Those that emanate from high school with technical education represent a percentage of 43, 90%.
– All of them declare that they have been trained in the use of the English language and they evaluate their knowledge as on average level in a percentage of 60%.
– Only a percentage that reaches the 17% is aware of a second foreign language, mainly German also reported to be used at an average level.
– The students that participated in the research have completed the compulsory educational travels,

thus they have completed their practical education, therefore, it is agreed that they have a completed aspect with regard to their education as well as the profession they will follow upon successful graduation.
– In the submitted questions to the students referring to the level of education, a percentage of 78% declared satisfied while a percentage of 48% evaluated the level of education as very good.

3 CONCERNED BASED ADOPTION MODEL, (C.B.A.M)

The Concerns-Based Adoption Model- (CBAM) mainly deals with the effort of description, measurement, and explanation of process of change in education.

The CBAM, is experienced both the instructors and their students that try to apply and follow new procedures in teaching and studying as well as the implemented educational material and practices. With the assistance of figure 1, it is shown that CBAM is a framework designed to help change facilitators indentify the special needs of individuals involved in the change process and address those needs appropriately based on the information gathered through the model diagnostic dimensions.

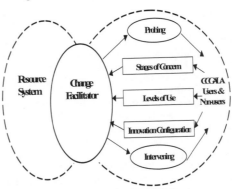

Figure 1. Concern Based Adoption Model

With those diagnostic data, the instructor could then develop a policy for any required interventions in order to facilitate the change effort. Together the SoC and LoU provide a powerful description of the dynamics of an individual involved in change; one dimension focusing on feelings, the other on behaviorural patterns.

4 STAGE OF CONCERN

Stages of Concern provide a framework from which to understand the personal side of the change process. According to this we assign different priorities and level of interest the things we perceive, individually and in various combinations, but most of the time we have little or no interest in most stimuli. Concerns are an important dimension in working with individuals involved in a change process like the simulation course in the maritime education.

Seven Stages of Concern about an innovation have been identified (see figure 2).They are called stages because usually there is a development movement through them. The participants of a simulator course may experience a certain type of concern although in the process they may experience another kind of concern. Another type of concern may emerge. The Stage of Concern about a simulator course appears progressively from little or no concern, to concern about the task of adopting by simulator, and finally to concern about the whole impact of the simulator course. The Stage of Concern Questionnaire (SoCQ) is the primary tool for determining in which stage the individual is.

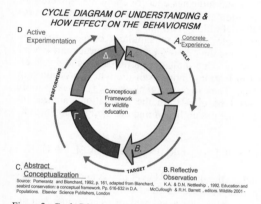

Figure 2 : Cycle Diagram of understanding

5 THE LEVEL OF USE

The levels of use represent models of behavior which are classified in eight separate categories. The focus is on what the individual is doing or not doing. Those behavior models mainly focus on the actions of individuals which have completed the course regardless if the outcome is successful or not (see table 2). Each model is recorded and analyzed as well as a series of individual reactions that are also connected with the particular behavior. The result is drawn on a table 5 that consists of the levels of use along with the seven categories which we are examining and the predetermined behaviors

6 METHODOLOGY-ANALYSIS

The methodology utilized in this research is the *focused interview*. A typical questionnaire contains questions in which we usually anticipate answers that are related to questions asked during each structure of object.

6.1 *MEASURING THE STAGE OF CONCERN*

The Stage of Concern Questionnaire (SoCQ) are developed in order to measure the concern of the individual. The data analyses were based on the following:

1 Determine the highest perceived concern for each participant, along with one or two also high concerns. The remaining stages are characterized by default, of lower concern.

2 SoCQ percentile stage scores (seven plus total)

The validity of the stage of concern questionnaire is investigated by examining the highest and the lowest scores at each stage separately and related to one or another stage. In the duration of the investigation all the other variables as concerns theory seizes are taken into consideration.

The participants in this study used a 0-7 scale to respond each item. The highest response indicated that the person considered items in its scale; the sum of the scale scores constituted the total score. Examining both the highest and the second highest a more detailed interpretation is possible. Analyzing the complete profiles allows us the most sensitive interpretations of responders.

6.2 *MEASURING LEVEL OF USE*

As a definition, the Levels of Use are the sequence a user which follows during his progression while using an innovation. It is addressed to the adapted methodology by the user while he gains his confidence with his developed skills in order to ascertain the use of the educational innovation. Following the same logic, however, a different individual may remain unchangeable during the duration of process of change.

The adopted methodology is outlined in the following 3 points:

– The questionnaire developers investigate the validity of the LoU by examining the relation of scores on the seven stages scales amongst each other as well as amongst variables suggested by theory of Level of use.

– The behaviors that are described at each intersection in the LoU Chart are derived from combining the described of a category of this level. Overall the Lou chart is describe and reliably measured. Each level of use represents a different approach in using a simulator.

- Each LoU is described in terms of the types of behaviors represented in the intersections in the chart of each category with a particular Level.

Table No 1 shows the levels of use according to model C.B.A.M. The individuals that participate in the educational innovation take for granted that they function in higher level than the level of routine IVA, so that the innovation is maintained and its use is adopted.

Table 1. STAGE OF CONCERN

Stage of Concern	Expression of Concern
6. Refocusing	I have some ideas about something that would work even better.
5. Collaboration	How can I relate what I am doing to what others are doing?
4. Consequence	How is my use affecting learners? How can I refine it to have more impact?
3. Management	I seem to be spending all my time getting materials ready.
2. Personal	How will using it affect me?
1. Informational	I would like to know more about it.
0. Awareness	I am not concerned about it.

Table 2: The level of use of the innovation
Levels of Use of the Innovation: Typical Behaviors

Levels of Use	Behavioral Indicators of Level
VI. Renewal	The user is seeking more effective alternatives to the established use of the innovation.
V. Integration	The user is making deliberate efforts to coordinate with others in using the innovation.
IVB. Refinement	The user is making changes to increase outcomes.
IVA. Routine	The user is making few or no changes and has an established pattern of use.
III. Mechanical	The user is making changes to better organize use of the innovation.
II. Preparation	The user has definite plans to begin using the innovation.
0I. Orientation	The user is taking the initiative to learn more about the innovation.
0 . Non-Use	The user has no interest, is taking no action.

From *Taking Charge of Change* by Shirley M. Hord, William L. Rutherford, Leslie Huling-Austin, and Gene E. Hall, 1987. Published by the Association for Supervision and Curriculum Development (703) 549-9110 Reprinted with permission.

Table 3: Result Table of stage of concern

Stage	Percent %	Group
0 Awareness	12.75%	SELF 39,25%
1 Informational	13.79%	
2 Personal	12.71%	
3 Management	11.42%	TASK 23,92%
4 Consequence	12.50%	
5 Collaboration	13.93%	PERFORMING 27,43%
6 Refocusing	13.50%	

In the case of Levels of Use it is assumed that the inquired person has a good comprehension of the theoretical frame Level of Use and then direct himself accordingly and proceeds by answering the questions. Even if the questionnaire is drawn with various types of discriminations so that it facilitates the user but also us, a lot of points exist where discrimination between the formal responses is sensitive, thus the inquired person will have to reply in a consistent and reliable way in order to be effectively evaluated.

In order to access and mark the usefulness of Engine Room Simulation, always according to the theory of Level of Use the behaviors of individuals were developed and delimited in the seven categories (see Table 4).

7 RESULT ACCORDING THE CONCERN

Towards classified the percentile result which obtained and following the theory of high and low stage of concern table 3 coming up. The results which can be extracted about the concern of participants according to the highest percentiles is that 13.95% indicates a great concern about collaboration between the participant and the relation which has been developed during the simulation courses. The interpretation for this result can be considered very logical, if we compare it with the real work environment in the engine room of a ship. The second highest percentage is 13,79% and indicates concern about the collecting of information and knowledge for the simulation course and the definition of all this process.

In whole process of simulator courses the way which cope the 13.50% present of participant is very interesting. That indicates participants who are interested in learning more from the whole procedure of the task. They focus on exploring ways to reap more universal benefits from the simulation courses, including the possibility of making major changes to it.

8 RESULT ACCORDING THE LEVEL OF USE

Based on the theory of Level of use and the response of participants the result of our research is classified between level of use and categories coordinates at the table No4. The highest percentage is in the category of sharing and at the category of assessing which is classified at the level of use V (integration).

Integration indicates that our participants combining their own efforts to use the simulation course with the related activities of colleagues to achieve a collective impact for their own common sphere of influence.

As for the category of assessing, participants indicate that they appraise collaborative use of the innovation in terms to increasing his own outcomes and strengths and weakness as the integrated effort.

At the categories of Status Reporting and at the Acquiring information the users indicated Level of Use IVA Routine: that means they determine the use of simulator and few changes take place in ongoing use.

The weaknesses of the users indentified on the knowledge, planning and performing category. Actually in this category the instructor must be given more attention in order to enhance the output of simulator course.

Table 4: Result table for Level of Use & categories

LEVEL OF USE		CATEGORIES						
LEVEL	CHARACTERISTIC	KNOWLEDGE	ACQUIRING INFORMATION	SHARING	ASSESSING	PLANNING	STATUS REPORTING	PERFORMING
VI	RENEWAL							
V	INTEGRATION			LoU V	LoU V			
IVB	REFINEMENT							
IVA	ROUTINE		LoU IVA				LoU IVA	
III	MECHANICAL USE	LoU III						
II	PREPERATION					LoU II		
I	ORIENTESION							LoU I
0	NONUSE							

9 PEDAGOGICAL VIEW

From pedagogical view the simulator courses can be used under cooperation of cognitive theories based on virtual reality learning. As mentioned at the beginning of our research many theories can be adapted in order to enrich the whole procedure. The final choice of correct theory must be a combination of the simulation task and the following correct theory. The correct approach of pedagogical theory during the course can be defined from the category of knowledge and performing at level of use. During our research we investigate and compare the closest pedagological theory which can be adapted to simulation course the result can be found in table 5. The main purpose of adoption of the pedagogical theory in simulation course is to engage the participant in the role of transferring and generating cognitive inside from the hall procedure of simulation curses as much as possible. In an effort to indentify in which stage of learning our participant are; we adapted the theory learning theory of Kolb. According to this theory typically expressed as four-stage cycle of learning,

1 Concrete Experience - (CE)
2 Reflective Observation - (RO)
3 Abstract Conceptualization - (AC)
4 Active Experimentation - (AE)

Furthermore we are trying to assign the results from the stage of concern with the ones from the stage of learning of Kolb theory. As the theory of concern says we categorized the seven stages of concern into three groups (see table 4). Those groups are: from stage 0 to stage 2 called Self Group. Stage 3 and 4 called Target group. Stage 5 and 6 called Performance group. We summarized the percentile score for each group and the result is for Self concern group 39.25%, for Target concern group 23.92%, and for group of performance 27.43%.

Finally we pair the data from stage of concern with the stage of learning (see figure 2) and the findings are that our participants in 39.25% are between the concrete experience and the observations. The 23,92% percent is between reflective observation and abstract conceptualization. Respectively the 27,43% are between the abstract conceptualization and active experimentation.

10 CONCLUSIONS

The conclusion of the research is the fact that the surveyed students experienced and accepted the introduction of simulation in training positively by expressing an interest to learn more about the application. One of the most important outputs, is the expression of strong concern on the basis of the developed collaboration amongst students, during the simulation.

This findings probably are the main outcome of our research given that:

1 The simulator course works like knowledge accelerator and transfer knowledge.
2 The purpose of the engine simulator is to duplicate activities carried out in the engine rooms of ships.
3 In the engine rooms of ships there should be established and imposed conditions of teamwork and cooperation, rather than surface level and should actually include exchange of views and knowledge amongst the crew.
4 27.43% participants fell comfortable with the simulation course and they are ready to apply their knowledge in order to learn by trial and error and ensure their thoughts about the operation of engine room.

It seems that the students are fully aware of the role and importance of simulation and begin to seek partnerships and relationships that would assist them to cope with the theoretical exercises that might also experience the simulator in practice. In conclusion, the findings that the instructors received on the reactions and concerns of students on the receptivity of the implementation of the simulator as part of their basic education could be positively described.

Also encouraging data for the full and smooth acceptance of simulation in education process have been addressed.
- At the end of the course the 39,25% of participants are between of concrete experience and observation which means they don't realize what exactly happened or they don't have enough time to react .
- Not given or there are not enough incentives from the instructor to the participants for planning or to performing during the course.

The participants indicate a low concern to organizing, managing and scheduling the simulation course.

11 THE EFFECT OF OUR RESULT

- The instructor decides to corporate more closely with the manufacture company to readjust the simulation course in the specific needs of student.
- The syllabus of Academy reorganized according to the needs of simulation course.
- The instructor decide to devote much more time at the stage of debriefing and more general to inform the participant for the main purpose of simulation course.

It is the first time that the C.B.A.M. "framework" is used as a tool to measure the outcome of the application of marine simulation in a merchant marine Academy. It was anticipated to provide an output related to the concerns and aptitude not only from the users but also the instructors. It could be used as a tool to measure the effectiveness between the investment of simulation and the knowledge which the students acquire.

12 FUTURE RESEARCH

This paper is an initial part of our research. It is in our future plan to apply the CBAM model in as much as possible maritime simulators applications, in Europe but further more in Asia. Our main proposal is to create a flexible tool which can be adapted in each maritime simulations course in order to determine the effects of the courses to the seafarers. From the other point of view, according to the response of the seafarers, it will determine the evaluation of the teaching and the pedagogical method which can be applied.

REFERENCES

Wickens C.D., Kramer A. (1985). Engineering Psychology. Annual Review of Psychology. Vol. 36: 307-348.

Newhouse C.P. (2001). Applying the Concerns-based Adoption Model to research on computers in classrooms. Journal of Research on Technology in Education, 33(5).

George A.A., Hall G.E., Stiegelbauer S.M. (2006). Measuring Implementation in schools: The Stage of Concern Questionnaire. Austin, Texas, SEDL.

Hall G.E., Dirksen D.J., George A.A. (2006). Measuring Implementation in scholls: Level of Use. Austin, Texas: SEDL.

Willis J. (1992). Technology diffusion in the soft disciplines: Using Social Technology to support information technology. Texas.

McKinnon D.H., Nolan P.C.J. (1989). Using computers in education: A concerns based approach to professional development for teacher. Australian Journal of Education Technology, No. 5, May.

Petruzella F.D. (1995). Industrial Electronics. McGraw-Hill.

Anderson S.E. (1997). Understanding Teacher Change: Revisiting The Concerns Based Adoption Model. *Curriculum Inquiry, 27*(3), 331-367.

Hord S.M., Stiegelbauer S.M., Hall G.E., George A.A. (2006). Measuring Implementation in scools: Innoviation Configurations. Austin, Texas: SEDL.

STCW Convention Resolutions of the 1995 conference. (1996). International Maritime Organization, London.

Avouris N., K. C. Issues Evaluation of Images collaborative environment. Wesley Publications

Kokotos D.H. (2007). Virtual environments. Stamoulis Publications, Athens.

Radar and Navigational Equipments

Reproduced with permission of the copyright owner. Further reproduction prohibited without permission.

21. Impact of Internal and External Interferences on the Performance of a FMCW Radar

P. Paprocki
Telecommunications Research Institute, Gdansk, Poland

ABSTRACT: Although FMCW technology has become by now mature, there are still some unsolved problems left. First group of them are noise and distortion from own transmitter or receiver. They are usually caused by DDS (Direct Digital Synthesis) and power converters performance. They appear on radar display as circles or arcs because of their isotropic properties. Second group are interferences from external world like beams of pulsed radars working in the same frequency band, reflections from close buildings and installations or sea clutter. Apart from creating artificial echoes of target they can cause degradation of radar tracking system. This paper evaluates influence of internal and external sources of interferences on the FMCW radar performance. DSP (Digital Signal Processing) algorithms which optimize detection of FMCW radar were presented. Means of suppressing analog interferences by digital techniques were shown as well.

1 INTRODUCTION

There are two basic kinds of interferences which decrease a performance of FMCW radar. The first kind are noise and distortion of the own radar, second one are interferences coming from nearby buildings, sea clutter and other electronic equipment working in the same frequency band. There are some ways to couple with such a problem. Most of them are based on digital signal processing algorithms implemented in DSP processor code. Developing of CFAR (Constant False Alarm Ratio) detector looks as the best way of suppressing analog interferences. The radar works in X band. In DPU (Digital Processing Unit) 8192 point real FFT was applied as well as binary integration, correlation and a few kinds of CFAR including clutter-map. The antenna and transceiver of FMCW radar are shown in the figure 1.

Figure 1. FMCW radar

2 INTERNAL INTERFERENCES

Although manufacturers of radars try to achieve perfect spectrum performance, there is always small leakage of interferences coming from the transmitter and the receiver. Some unwanted products of digital synthesis and power converters appear in the IF (Intermediate Frequency) signal. Although the signal sampled in DPU (Digital Processing Unit) is band limited, some of the mentioned products remain in data vector after FFT (Fast Fourier Transform) processing. Having a level greater than background noises, they are detected and displayed as artificial echoes. Because interferences appear in the fixed frequency, they are seen as a circle or a part of the circle on the radar display (Fig. 2). Such kind of artifact can mislead radar operator and tracking system. Echoes of ships which are close to the circle would be probably undetected. Tracking of the object moving trough the interference area is frequently lost. There are some simple ways of suppressing such interferences. Firstly, we can increase the detection level. Unfortunately, in this case we reduce the detection probability of others echoes as well. It is also possibility to change carrier frequency of transceiver. Although the interference (circle) vanishes, another one often appears nearby.

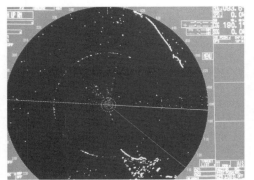

Figure 2. Circle-shaped interferences as a result of the own transmitter noise and a radial interference from the pulsed radar beam.

To analyze the problem from DSP point of view we should consider how CFAR algorithm works [3].

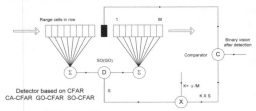

Figure 3. CFAR algorithm concept

The range-cell under test is compared with its neighborhood. If the value of the signal in the cell is greater then the average value of the neighborhood, detection occurs. Specific interference exists only in a few range cells in a row. If we found the right cell number, we could increase the CFAR level locally. Even though real echoes in those range-cells are suppressed as well, other range-cells are unaffected. This simple approach suppresses the interference just before detection. The results of applying such approach is shown in figure 4. The circle disappears, detection of nearby echoes is still possible.

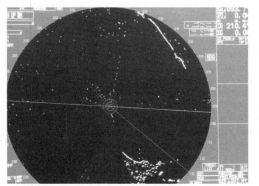

Figure 4. Circle-shaped interferences suppressed digital algorithm.

3 EXTERNAL INTERFERENCES

In order to limit the influence of other electronic devices working in the same band of frequency FMCW signal should be conditioned. The typical example is a problem with beams of the pulsed radars which appear on the display as a radial row of dots (Fig. 2). High level of such a radar signal masks the nearby echoes. Data conditioning algorithm limits the level of time-domain data to average signal level in the sweep. It results in limiting unwanted products after FFT processing. Second source of external interferences are reflections coming from nearby buildings and sea clutter. Distinguishing sea clutter with real objects is often problematic. In any case some kind of filtering is needed [1]. Possible solution is a filter based on clutter-map implemented as a type of CFAR [2].

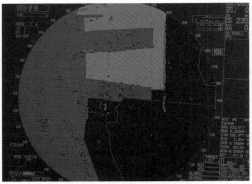

Figure 5. Gdynia harbor 0.25Nm range GO-CFAR

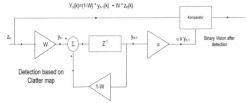

Figure 6. Clutter map algorithm concept

Clutter-map estimate caries information about the past and present values of particular range-angular cell. The more up-to-date the data is, the greater contribution in estimate it has. The state of the estimate is continuous up-dated in RAM memory. The value of the estimate can be written as:

$$Y_n(k) = (1-W) * y_{n-1}(k) + W * Z_n(k)$$

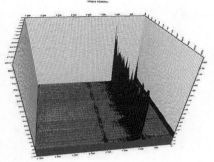

Figure 7. Clutter-map estimate of 32 sweeps in a row (Y-axis).

To store the estimate N x M x 2 bytes of RAM are used (N-number of range cells, M- number of sweeps per antenna rotation). For N=4096 and 3000 sweeps per rotation 24 MB of RAM is occupied only by clutter-map estimate. The need to access to a large and relatively slow external SDRAM, which deteriorates processing speed is the main disadvantage of such kind of CFAR.

Detection probability is given by the following expressions [4]:

$$P_D = \frac{1}{\prod_{m=0}^{M} [1 + \alpha_D W (1 - W)^m]}$$

$$\alpha_D = \frac{\alpha}{1 + SNR}$$

$$M \to \infty$$

The W and α coefficients should be optimized in order to achieve desired probability of detection for maneuvering objects.

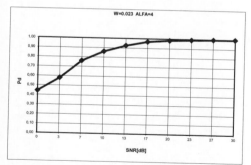

Figure 8. Probability of detection diagram.
W=0.023, K=4.

There are a few characteristic effects of clutter-map algorithm. Any fixed echoes coming from nearby buildings, boardwalks and buoys disappear. In some applications radar echoes of such objects

are not needed. Moreover, they are still visible on electronic chart. Sea clutter is suppressed by such kind of detector as well. Figure 5 shows the radar display of Gdynia harbor obtained using GO (Greatest Of)-CFAR type. There are many items redundant including unwanted echoes from nearby port facilities and from the coast line. Results of the detection based on clutter-map are shown in the figure 9. Only maneuvering objects are still visible. Such kind of data processing gives excellent detection performance in case when a small object is moving along boardwalks or quay. In the classical CFAR such object would be overwhelmed by the strong echoes from large fixed objects. Another advantage of clutter-map in context of the tracking system, is a limitation of data stream, because plots which are sent from extractor are preliminary filtered.

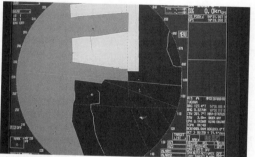

Figure 9. Gdynia harbor 0.25Nm range CFAR based on clutter-map.

4 CONCLUSIONS

In this paper an approach to improve FMCW radar detection was presented. All measures of suppressing internal and external interferences were based on different kind of CFAR. It was shown how to eliminate false, artificial echoes which may mislead the operator or the tracking system. Some advantages of detection based on clutter-map were considered as well.

REFERENCES

[1] Keith D. Ward, Robert J.A. Tough and Simon Watts. "Sea Clutter: Scattering, the K Distribution and Radar Performance" . Institution pf Engineering and Technology, London 2006

[2] Conte E. "Clutter-Map Detection for Range –Spread Targets un Non-Gaussian Clutter." IEEE Transaction on aerospace and electronic systems Vol. 33 April 1997

[3] Khalighi M.A. "Adaptive CFAR Processor For Nonhomogeneous Environments" IEEE Transaction on aerospace and electronic systems Vol. 36 July 2000

[4] Nadav Levanon "Radar Principles" Wiley Interscience Publication 1988

22. Fusion of Data Received from AIS and FMCW and Pulse Radar - Results of Performance Tests Conducted Using Hydrographical Vessels "Tukana" and "Zodiak"

A. Król, T. Stupak & R. Wawruch
Gdynia MaritimeUniversity, Gdynia, Poland

M. Kwiatkowski, P. Paprocki & J. Popik
Telecommunication Research Institute Ltd., Gdańsk Division, Poland

ABSTRACT: Paper presents results of performance tests of the Integrated Vessel Traffic Control System realizing fusion of data received from shore based station of the Automatic Identification System (AIS) and pulse and Frequency Modulated Continuous Wave (FMCW) radars and presenting information on Electronic Navigational Chart issued by the Polish National Hydrographical Service – Hydrographical Office of the Polish Navy. Tests were conducted in real sea conditions using hydrographical vessels "Tucana" and "Zodiak" owned by the Maritime Office in Gdynia.

1 INTRODUCTION

Integrated Vessel Traffic Control System realizing fusion of data received from shore based station of the Automatic Identification System (AIS) and pulse and Frequency Modulated Continuous Wave (FMCW) radars and presenting information on Electronic Navigational Chart issued by the Polish National Hydrographical Service – Hydrographical Office of the Polish Navy is described in other paper presented on this conference. Described system was designed, built and tested in the scope of research work financed by the Polish Ministry of Science and Higher Education as developmental project No OR00002606 from the means for science in 2008-2010 years.

This paper described results of the exploitation test of the constructed system conducted in real sea conditions using hydrographical vessels "Tukana" and "Zodiak" owned by the Maritime Office in Gdynia.

2 DESCRIPTION OF THE MEASUREMENTS

Measurements were conducted on 30th of September and 24th of November 2010 using:
- Installed on shore in the radar laboratory of the Gdynia Maritime University:
  - Pulse radar Raytheon NSC 34;
  - FM-CW radar built by the Przemysłowy Instytut Telekomunikacji S.A.; and
  - Class A ship borne AIS type R4 produced by SAAB.
- Hydrographical vessel "Zodiak" equipped with:

- Receiver GPS RTK Trimble R7 L1L2 using during the tests GPS reference station ID 745 situated in Gdynia; and
- Ship borne AIS R4 produced by SAAB.
- Hydrographical vessel "Tucana" equipped with:
- Receiver GPS RTK R7 using during the tests the same GPS reference station; and
- Ship borne AIS R4 produced by SAAB too.

Basic parameters of utilized ships and radars are described in Tables 1-3. Vessels are presented in Figures 1 & 2.

Table 1. Basic parameters of the hydrographical vessels "Zodiak" and "Tucana"

Ship's parameter	Value	
	Zodiak	Tukana
Displacement	751	71
Length	61.3 m	23.0 m
Breadth	10.8 m	5.8 m
Mean draught	3.3 m	2.2 m
Power	2 x 706 kW	2 x 280 kW
Maximum speed	14.0 knots	12.0 knots
Height of the radar scanner above sea level	16 m	6.5 m

During the measurements were good weather conditions without rainfalls. There were sea waves 0.5-1m high.

The main goal of the test was assessment of the accuracy of information about ship's position, course over ground (COG) and speed over ground (SOG) received from both shore based radars and AIS before and after their fusion in comparison with values of these parameters indicated by ship borne DGPS receiver. Due to that there were automatically registered positions, COG and SOG of the vessel indicated by:

Table 2. Basic parameters of the radar Raytheon NSC 34

Parameter	Value
Output power	25 kW
Carrier frequency	9410±30 MHz
Range	0.25; 0.5; 0.75; 1.5; 3; 6; 12; 24; 48; 96 NM
Pulse length	0.06 μs, 0.25 μs, 0.5 μs, 1.0 μs,
Pulse repetition freq.	3600 Hz, 1800 Hz, 900 Hz
IF bandwidth	20 MHz, 6 MHz
Antenna length	2.1336 m
Beam width horizontal/vertical	1.0°/23° dB
Polarisation	Horizontal
Gain	29 dB
Rotation speed min/max.	22/26 rpm
Display size	28 inch
Resolution	1600 × 1200 pixels
Acquisition	automatic up to 40 targets
Tracking	automatic of all acquired target
Range accuracy	0.3% of selected range or 6.4 m (whichever is greater)
Angle resolution	0.3°
Bearing accuracy	1.0°

Table 3. Basic parameters of the FM-CW radar

Parameter	Value
Output power	1mW-2W (switched)
Carrier frequency	9.3 – 9.5 GHz
Frequency deviation	54 MHz at 6 NM
switched according to	27 MHz at 12 NM
the required scale range:	13.5 MHz at 24 NM
Range scales	0.25 NM – 48 NM
Modulation	DDS based linear FMCW
Sweep repetition period	1 ms
IF bandwidth	4 MHz
Frequency curve slope of IF amplifier	6 dB/oct; 12 dB/oct; 18 dB/oct
Antenna length	3.6 m
Beam width horizontal/vertical	0.70°/22° dB
Polarisation	Horizontal
Gain	32 dB
Rotation speed min/max	12/30 rpm
FFT signal processing	8192-points FFT
Sampling frequency	8 MHz
Display size	22 inch
Resolution	1280 × 1024 pixels
Acquisition	automatic up to 100 targets
Tracking	automatic of all acquired targets
Range accuracy	1% of selected range or 50 m (whichever is greater)
Angle resolution	0.1°
Bearing accuracy	0.7°

Figure 1. Hydrographical vessel "Zodiak"

Figure 2. Hydrographical vessel "Tucana"

– GPS RTK receivers onboard ships;
– AIS and radars Raytheon NSC 34 and FM-CW installed onshore in radar laboratory; and
– Display unit of the integrated system after fusion of data from above mentioned sensors.

Figure 3 presents complete track of the ship "Zodiak". Dots indicate positions in ten minutes time intervals.

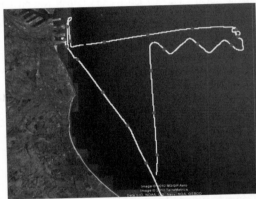

Figure 3. Complete track of the ship "Zodiak" during the measurements showing its positions received from onboard DGPS MX 420 receiver.

In figure 11 is presented part of the "Zodiak" track utilised for analysis presented in further part of this paper.

Figure 4 shows track of the vessel "Tukana" during the measurements received from onboard GPS RTK receiver and from shore based radar Raytheon.

3 RESULTS OF THE MEASUREMENTS

3.1 *Accuracies of the data available on shore from AIS and radars*

Tested integrated system performs fusion of data received from two connected radars and AIS. Due to that one of the main goals of the described tests were measurements of the accuracies of data available from shore based AIS station and radars and computation of their errors by comparison with data registered in onboard GPS receiver for the same moments of time.

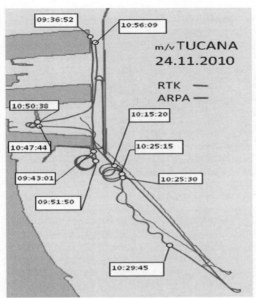

Figure 4. Track of the vessel "Tucana" during the measurements received from onboard GPS RTK receiver and from shore based radar Raytheon.

Figure 5 presents differences during the test between ship's positions, course over ground (COG) and speed over ground (SOG) available for the same moments of time onshore from AIS station and onboard "Zodiak" from DGPS receiver. They were accounted as differences between data: last received from onshore AIS and presented by onboard DGPS receiver. Errors arise due to the time differences between receiving by AIS data from ship's GPS receiver and its transmission according to the SOTDMA time schedule mainly. Additionally not all ship's transmissions were received by the shore AIS station. For ship's speed during the test (approximately 4.6 m/s) vessel's way between two consecutive AIS transmission was equal to 45 m (for passages with steady courses) or approximately 15 m (for course and/or speed alteration). Any fault in receiving AIS message causes errors directly proportional to this way.

Received results are compliant to conclusions done on the base of measurements of the AIS data accuracies conducted on the ship "Dar Młodzieży" and described in [1].

Radars begun to track echo of „Zodiak" after leaving by this ship port in Gdańsk and lost vessel at the entrance to the port in Gdynia when they connected its target with echo from coastline. Tracking was resumed after leaving the port and continued during ship's passage back to the Gdańsk.

Figure 6 presents differences between values of courses speeds and positions indicated onshore by ARPA Raytheon NSC 34 and by onboard DGPS receiver.

Figure 7 shows accuracies of ship's positions received from FM-CW radar during the test conducted with vessel "Zodiak" and calculated in relation to its AIS positions. Errors distribution of these positions indicated incompatibility of cartographical coordinate systems used for calculation of ship's positions transmitted by AIS and received on the base of radar measurements. It was eliminated by corrections introduced to the radar coordinates (0.0049' to the North and 0.0016' to the West). Corrected radar data was used for its fusion with information received through AIS system.

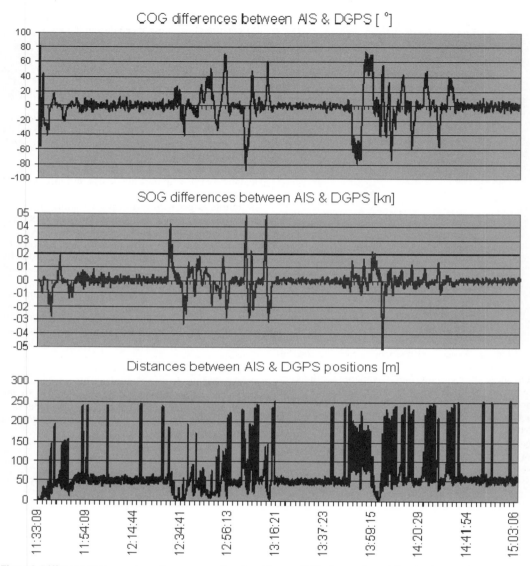

Figure 5. Differences between values of courses, speeds and positions available from onshore AIS and DGPS receiver onboard the ship

162

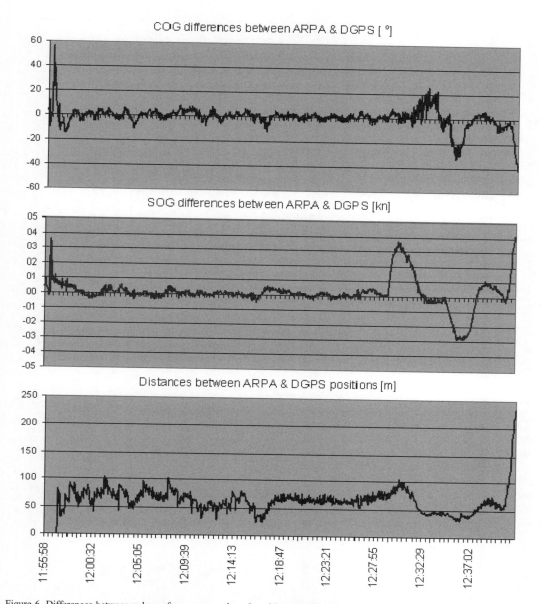

Figure 6. Differences between values of courses speeds and positions available from onshore radar Raytheon and DGPS receiver onboard the ship

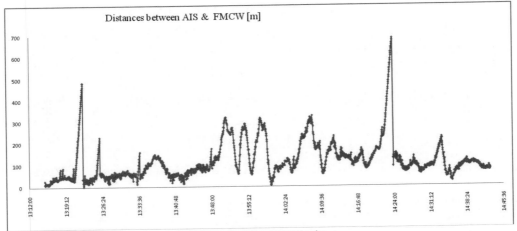

Figure 7. Accuracy of positions of vessel "Zodiak" received from FMCW radar

3.2 *Data fusion*

Data after its fusion is presented in Figures 8-11.

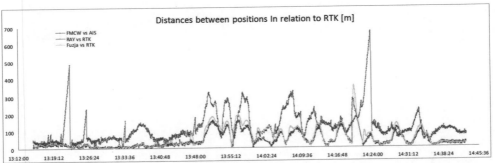

Figure 8. Distances between ship's positions indicated by particular radars and display unit of the integrated system (after data fusion) and AIS or GPS receiver

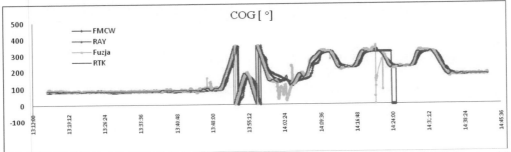

Figure 9. Information about ship's course over ground (COG) indicated by onboard DGPS receiver (RTK) shore based FMCW (FMCW) and pulse (RAY) radars and integrated system (after data fusion - Fuzja)

164

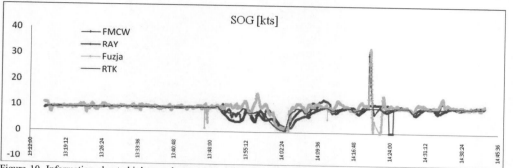

Figure 10. Information about ship's speed over ground (SOG) indicated by onboard DGPS receiver (RTK) shore based FMCW (FMCW) and pulse (RAY) radars and integrated system (after data fusion – Fuzja)

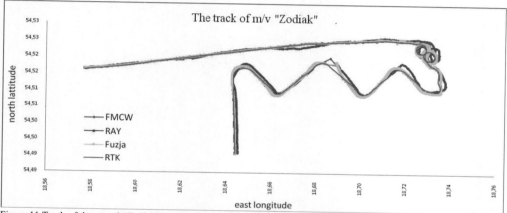

Figure 11.Track of the vessel "Zodiak" during tests plotted on the base of data available from onboard DGPS receiver (RTK) and shore based FMCW (FMCW) and pulse (RAY) radars and display unit of the integrated system after data fusion (Fuzja)

4 CONCLUSIONS

According to recommendation of the International Maritime Organisation (IMO) integrated navigational system (INS) shall, after checking that information received from AIS and radar tracking facilities concerns the same object, present for this object AIS data only. Conducted tests proved that AIS data may present inexact information about present vector of object's movement. Main reasons of this fault are problems with receiving AIS messages without time delay. Fusion of AIS and radar data as presented in this paper may be helpful in solving indicated problem. Additionally fusion of radar data increase reliability of radar data and its accuracy.

REFERENCES

[1] Król A., Stupak T: Dokładność rejestracji danych pozycyjnych statków w systemie nawigacji zintegrowanej. XIV International Conference Transcomp 2010, Logistyka 6/2010 str. 1675 – 1681

23. Statistical Analysis of Simulated Radar Target's Movement for the Needs of Multiple Model Tracking Filter

W. Kazimierski
Maritime University of Szczecin, Poland

ABSTRACT: The quality of radar target tracking has a great impact on navigational safety at sea. There are many tracking filters used in maritime radars. Large group of them are multiple model filters in which different filter parameters are used for different states (models) of vessel movement. One of possible filter is multiple model neural filter based on General Regression Neural Network. Tuning of such filter means to adjust its parameters for a suitable target movement model. This paper shows the results of an experiment aiming at determining such models based on statistical analysis of target's movement parameters. The research has been carried out with PC-based simulator in which typical radar measuring errors were implemented. Different manoeuvres of targets have been examined. Based on this, the possibility of movement models description has been stated as conclusion.

1 INTRODUCTION

Radar target tracking is one of the key issue influencing navigational safety of vessels at sea. For several dozen of years radar has been present on board of the ships establishing its position as a very important device on the bridge. It has been commonly used for observation of navigational and collision situation in the vicinity of own vessel. In the restricted visibility it is even the basic source of information, while remaining additional and complementary (to visual observation) source in good visibility.

1.1 *Tracking of maneuvering targets in radars*

Radar's functionality increased rapidly after implementing of target tracking facilities in ARPA systems. Since then it has become possible to support navigator's work by replacing manual plotting with automatic target tracking. The quality of tracking depends however on the implemented tracking algorithm. At the beginning relatively simple numerical algorithms, like α-β were used. In time those were replaced by more complex numerical algorithms based on statistical estimation, like Kalman Filter (Bole et al. 2005). It's main deficiency is the assumption of linear movement of the target, which leads to large errors and delays of tracking during target's and own ship's manoeuvres. These limitations are commonly known to the navigators. Various modifications of Kalman Filter (e.g. Extended

Kalman Filter, Unscented Kalman Filter) improved the quality of tracking significantly. The main goal was to include non-linear movement of maneuvering vessels into the algorithm. Thus better quality of tracking was achieved.

One of the possible solution of non-linearity problem is to create a few different filters for different motion stages (linear/ non-linear). This approach is called multiple-model filtering and is thoroughly examined for example in (Bar-Shalom & Li 1998).

Another possibility is to use typically non-linear methods for tracking, for example artificial intelligence. An interesting example might be Intelligent Kalman Filter presented in (Lee et al. 2006).

For several years the research focused on use of artificial neural networks in radar target tracking has been carried out in Maritime University of Szczecin. Particularly interesting results were obtained while using General Regression Neural Network (GRNN), which was presented for example during TransNav 2007 (Stateczny & Kazimierski 2007).

1.2 *Research project and paper scope*

The experience on target tracking with neural networks gained so far, resulted in preparing new research project in Maritime University of Szczecin, called *Elaborating of methods for radar tracking of maritime targets with the use of multiple model neural filtration*. The main goal of the project is to combine neural tracking filters with multiple model philosophy, traditionally used for numerical filters.

Different neural filters will be adjusted to track targets with different dynamics of movement. This means that, as the first stage of the project, several models of target's movement has to be declared. The aim of research presented in this paper was to perform statistical analysis of different target's movement to conclude on how to find these unique movement models.

2 GRNN FILTER FOR TRACKING IN MARINE RADARS

The filter proposed for radar target tracking and examined in presented research was based on General Regression Neural Network invented by D. F. Specht (Specht 1991), which is basically neural implementation of kernel regression algorithms presented in (Nadaraya 1964) or (Watson 1964). The structure of the network is strictly defined, but it needs some kind of adjusting to solve particular problem. This means mainly determining of input and output vectors, teaching sequence, radial neurons activation function and smoothing factor of it.

2.1 Tracking with GRNN

The concept of using GRNN to track radar targets in maritime navigational radars was shown in (Juszkiewicz & Stateczny 2000), (Stateczny & Kazimierski 2005) and (Kazimierski 2008). The filter proposed in these papers consists of two parallel GRNNs. One of them is to etimate Vx and the other Vy. For additional smoothing of signal, which means more stable vector of target on the radar screen, the second filtration stage, with another pair of the same networks is used. To ensure proper functioning of the filter, since the beginning of observation, the dynamic increase of number of radial neuron in hidden layer and elements of teaching sequence is introduced. Observed (measured) values of movement vectors are used as input and teaching values while estimated movement vector is the output. Movement vector is defined as (1).

$$V = \begin{bmatrix} V_x & V_y \end{bmatrix}^T \tag{1}$$

where V_x = speed vector over x axis, V_y = speed vector over y axis.

Both of the networks can be joined into one - more complex structure presented in the figure 1. Such a network has two basic parameters – the smoothing factor and the length of teaching sequence, usually both adjusted empirically.

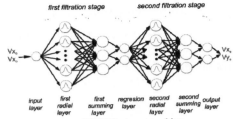

Figure 1. Two-stage GRNN for target tracking.

The smoothing factor determines the range of gaussian function in radial neurons and the teaching sequence determines how many observed vectors are included in estimating the state vector.

GRNN performs kernel regression, resulting in computing weighted average of teaching vectors. The weights are the values of Gaussian kernel function for the distances of input vector to teaching vector. Thus the estimation of movement vector is calculated according to following equation (Kazimierski 2008):

$$\begin{bmatrix} Vxe_i \\ Vye_i \end{bmatrix} = \begin{bmatrix} \dfrac{\sum_{i=1}^{n} Vxo_i \cdot e^{-\left(\frac{\|t-t_i\|}{2\sigma}\right)^2}}{\sum_{i=1}^{n} e^{-\left(\frac{\|t-t_i\|}{2\sigma}\right)^2}} \\ \dfrac{\sum_{i=1}^{n} Vyo_i \cdot e^{-\left(\frac{\|t-t_i\|}{2\sigma}\right)^2}}{\sum_{i=1}^{n} e^{-\left(\frac{\|t-t_i\|}{2\sigma}\right)^2}} \end{bmatrix} \tag{5}$$

where Vxe and Vye = estimated speed vector on axis x and y, Vxo and Vyo – observed speed vector on axis x and y, σ = smoothing factor of Gaussian kernel function, t = actual time step, t_i = former time steps.

2.2 Multiple model filtering

Multiple model approach is the development of so called decision based filters. The main idea is similar. The filter consists of a few elementary filters, each of them tuned to track target in unique movement stage, called model. They are running simultaneously. The final estimation can be a chosen output of one of elementary filters (in the decision based methods) or a combination of elementary estimates (in multiple model approach).

There are several particular algorithms of multiple model tracking, in which different interaction methods between elementary filters is used. Usually the probability of target being in each particular mode state is the criterion. Thus the estimated state vector is weighted average of elementary estimates. Fine description of most popular multiple model methods is given in (Bar-Shalom & Li 2001) and (Li & Jilkov 2005).

2.3 GRNN multiple model filter

The empirical research (Stateczny & Kazimierski 2006) or (Kazimierski 2007) has shown that different values of smoothing factor and of teaching length are needed in GRNN filter for different movement characteristics. For uniform motion – longer teaching sequences and bigger smoothing factors and for maneuvers shorter teaching sequences and smaller values of smoothing factor are expected. This gave the idea of creating multiple model neural filter, which can be implemented as decision based filter as well. A suitable patent application was issued. An example of such a filter is given in figure 2.

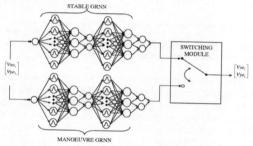

Figure 2. GRNN filter for target tracking.

Main problem in such an approach is to tune elementary filters for suitable movement model. This of course shows the need of defining such models.

3 NUMERICAL EXPERIMENT

The research based on simulation presented in this paper is just an initial phase and preparation for further parts in which real data will be involved. This time PC- based tracking radar simulator was used.

3.1 Experiment overview

The main goal of the experiment was to find any statistical dependency, that can be useful for defining tracked targets movement models. To ensure usefulness of experiment results for any tracking method, the unfiltered data were analyzed. These were obtained in the simulator by implementing suitable noise of measurements prior to any filtration.

To obtain statistical information, 100 Monte Carlo runs were performed for each research scenario. For each run, an average value and a standard deviation of ship's course, speed, Vx, Vy and increments of Vx and Vy as well as covariance between Vx and Vy were calculated. After the simulations, obtained

values were examined and analyzed with the use MS Excel

The research scenarios were planned in such a way to examine both uniform motion state and maneuver state.

3.1.1 Simulator description

The simulator used in research showed in this paper is a PC-based application, prepared by the author in MS Visual Studio.

The idea of radar target simulation used in the simulator derives from (Kantak et al. 1988) and is based on adding to non-cluttered measurement, process noise. Thus the position of simulated target is obtained. The noise is calculated as a product of maximum sensor noise and pseudo-random value. Start point of random numbers is changing, which allows carrying out Monte Carlo simulation.

Own ship movement is also simulated and typical errors of gyrocompass (0,5°) and log (0,05 kn) are included. The auto-correlation function factors were established based on (Stateczny et al. 1987).

The simulator has also other possibilities and functionalities, which were not used for the research for this paper, however they can be used for many other purposes.

3.1.2 Research scenarios

The idea of the research is to find different movement models based on statistical analysis of non-filtered target data. The research scenarios therefore include both - uniform target movement and maneuvers.

The first part of research focused on finding statistics for linear movement as the basis for comparison with maneuvering stages. Five different scenarios were examined for uniform movement. Initial situation was the same for each of them, except of course and speed values which differ for particular scenarios. The scenarios are described in Table 1.

Table 1. Scenarios for uniform movement

Scenario no	1	2	3	4	5
			Initial situation		
Bearing			030°		
Range			8 Nm		
Own ship course			000°		
Own ship speed			10 kn		
			Target movement parameters		
Target course [°]	135	180	270	135	135
Target speed [kn]	10	10	10	20	30

These research scenarios allowed to check the influence of various speed and courses on statistical dependences of target movement during steady movement.

The second part of research aimed at finding results during maneuver of target. The maneuver of course changing was examined as the most and advised in COLREG popular way of collision avoidance. The maneuver was applied with different rate of turn in different scenarios. As the same change of course was assumed, the maneuvers were lasting for different time in each scenario. The statistics were calculated only for the time during maneuver. Detail description of scenarios can be found in Table 2.

Table 2. Scenarios for maneuvering target

Scenario no	1	2	3	4	5
Initial situation					
Bearing	030°				
Range	8 Nm				
Own ship course	000°				
Own ship speed	10 kn				
Target course	135°				
Target speed	10 kn				
Course maneuver					
Course change	90° to starboard				
New course	225°				
Rate of turn [°/min]	10	20	30	40	50

Examining of the maneuvers with different dynamics allowed to answer the question if there is any statistic dependent of turn rate, which can become a basis for establishing movement models in multi model filter.

Each scenario covers 200 measurement steps, which means about 10 minutes of simulation time.

3.2 Results of experiment

The simulator used for experiment prepares the output statistics for each of 100 Monte Carlo runs in ASCII file. In the next step it was imported to MS Excel to prepare graphs and to perform further analysis. The results are divided into two parts – uniform motion and maneuver. The conclusions are stated for each part separately and then jointly for all simulations.

3.2.1 Uniform motion

The scenarios in which the course was different were analyzed together and the scenarios in which the speed was different were also analyzed jointly.

Figure 3 shows the standard deviation of Vx during simulation for each of 100 runs. Scenarios 1, 2 and 3 were included. It can be noticed, that the value of standard deviation does not vary significantly for the scenarios, although in case of scenario 3 the values of standard deviation is slightly bigger than in other scenarios.

Similar results were obtained for other measured values – standard deviation of Vy, course and speed. this can lead to the conclusion that standard deviation of movement vector parameters does not change

significantly in case of uniform movement with different courses.

In figure 4 the same standard deviation of Vx is presented but for the scenarios in which the target was moving uniformly but with different initial speed. Once again the value of standard deviation seems not to vary much in different scenarios.

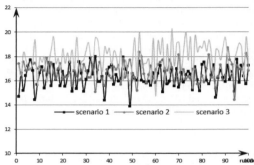

Figure 3. Standard deviation of Vx during 100 Monte Carlo runs for uniform motion of target with different course.

The values for scenario 5 in which the speed was the biggest are usually a bit smaller. As similar results were obtained for other parameters it can be concluded, that standard value of them does not change significantly for the uniform motion, even if the speed is different.

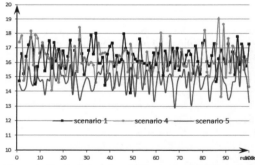

Figure 4. Standard deviation of Vx during 100 Monte Carlo runs for uniform motion of target with different speed

An interesting issue of statistical analysis of movement vector can be covariance of Vx and Vy vectors. It was measured as the covariance of random samples Vx and Vy. Figure 5 shows the average value of average covariance in each of Monte Carlo runs. It can be noticed that, although average value vary for different scenarios, the standard deviation remains on the same level. This means that covariance value vary in different scenarios, but it remains in the same "statistical frame" for all of them and it could not be easily stated which scenario is it, based only on the results.

The most important conclusion of the first part of experiment is that one movement model for non-maneuvering target is sufficient. Changes of course and speed do not influence significantly on statistical factors for movement vector parameters. The next question is if this is also true for maneuvering target.

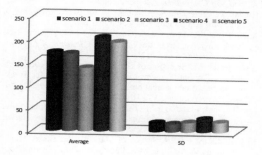

Figure 5. Average value and standard deviation for average covariance in 100 Monte Carlo runs for non-maneuvering target.

3.2.2 Maneuver

In the second part of research, maneuvering target was observed. Only course maneuver was implemented. Based on earlier works it was assumed, that conclusions for speed maneuver would be similar.

One maneuver was examined but with five different dynamics, represented by rate of turn.

Figure 6 contains four graphs in fact. Two of them present average values of Vx and Vy during maneuver and two other presents standard deviations of these. The values are presented for 100 Monte Carlo runs. Only two selected scenarios are presented on this figure, namely scenario 1 (rate of turn = 10°/min) and scenario 5 (rate of turn = 50°/min). Graph for these extreme values presents the nature of statistics sufficiently and adding other (middle valued) scenarios to the graph would only decreased its readability.

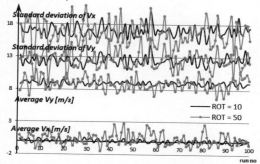

Figure 6. Average values and standard deviations of Vx and Vy in the scenarios with target course maneuver

It can be derived from figure 6, that average values of both Vx and Vy for different scenarios are more or less on the same level, however the variance of them is definitely bigger for more dynamic maneuvers. Similar observation can be made for standard deviation of Vx and Vy. Although in this case it is not so obvious, but larger spread of standard deviation values for maneuvers with bigger rate of turns can be noticed.

The conclusions derived from figure 6 should be confirmed with the analysis presented on figure 7.

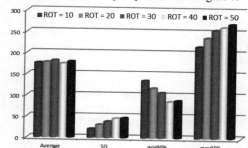

Figure 7. Statistics for covariance of Vx and Vy in 100 Monte Carlo runs.

It can be easily noticed, that average value of covariance remains on the same level for each scenario. What seems to be very interesting, the value of standard deviation is changing at the same time and the tendency is obvious. The faster the turn is (rate of turn is bigger), the bigger standard deviation – the covariance is more spread.

In the figure 7 the boundaries of 95% confidence levels are additionally shown. It was calculated based on average and standard deviation values. These two graphs visualize directly how the region of 95% confidence is enlarging with the increase of rate of turn value.

An important conclusion derives from figures 6 and 7, namely that dynamic of the maneuver can be noticed during statistical analysis. Especially the value of covariance of Vx and Vy can be very useful in determining maneuver rate.

3.3 Conclusions

To conclude jointly the research figures 5 and 7 shall be compared. The average value of calculated covariance is basically the same for non-maneuvering and for maneuvering targets. The standard deviation on the other hand is clearly increasing as the rate of turn is getting bigger. So for the steady motion standard deviation is small and for fast maneuvers is bigger.

Statistical analysis of movement vector parameters (course, speed, Vx, Vy) can also be used for differing steady motion from maneuvers, the analysis however is not so obvious.

This leads to a conclusion, that covariance between Vx and Vy is the best value to define movement models and the definition should be based on standard deviation analysis.

4 SUMMARY

The idea of building multiple model neural filter seems to be promising alternative for numerical filters in the light of earlier research.

Definition of movement models is of the key issue for this project. The paper presented the analysis of possibility of determining such models, based on statistic dependences. The results of the research showed that statistical analysis of covariance of movement vector elements (Vx, Vy) can be particularly useful for this purpose. It has been proven that standard deviation of such a covariance is increasing when the target is maneuvering faster.

The determination of particular models, based on standard deviation threshold, shall be the subject of empirical research. This will probably be one of the future steps for continuation of presented research. However prior to these more simulation research shall be conducted. These shall include especially the influence of own ship – target geometry for the statistics observations.

An important conclusion derived from research is also the fact that one movement model is sufficient for describing uniform motion, while for the maneuvers a few models shall be established. The number of them should be the result of empirical research.

Another problem will be to "translate" statistical model to GRNN, thus to adjust network parameters accordingly.

REFERENCES

Bar Shalom Y., Li X.R.: *Estimation and tracking: principles, techniques, and software*, YBS, Norwood, 1998.

Bar Shalom Y., Li X.R.: *Estimation with Applications to Tracking and Navigation: Theory Algorithms and Software*, John Wiley & Sons, Inc., NY USA, 2001

Bole A. G., Dineley W. O., Wall A.: *Radar and ARPA Manual*, Elsevier Science & Technology Book, 2005

Juszkiewicz W., Stateczny A., GRNN Cascade Neural Filter for Tracked Target Maneuver Estimation, *Neural Networks and Soft Computing*, Zakopane 2000

Kantak T., Stateczny A., Urbański J.: *Basis of automation of navigation* (in polish). AMW, Gdynia 1988.

Kazimierski W., Two – stage General Regression Neural Network for radar target tracking, *Polish Journal of Environmental Studies*, Vol. 17, No 3B, 2008.

Kazimierski W.: Selection of General Regression Neural Network's Training Sequence in the process of Target Tracking in Maritime Navigational Radars, *Polish Journal of Environmental Studies*, Vol 16A., 2007

Lee B.J, Park J.B., Joo Y.H., Jin S.H.: Intelligent Kalman Filter for tracking a manoeuvring target, *IEE Proceedings: Radar, Sonar and Navigation Vol. 153*, IET, Stevenage UK, 2006.

Li X.R, Jilkov V.P.: A Survey of Maneuvering Target Tracking—Part V: Multiple-Model Methods, *IEEE Transactions on Aerospace and Electronic Eystems*, Vol. 41, 2005.

Nadaraya E. A.: On estimating regression. *Theory of Probab. Applicat.*, vol. 9, pp. 141–142, 1964.

Specht D. F.: A General Regression Neural Network, *IEEE Transactions on Neural Network*, Vol. 2, No. 6, 1991.

Stateczny A., Felski A., Krotowicz M.: Generating of correlated measurements in navigational research (in polish), *Biuletyn WAT*, 1987

Stateczny A., Kazimierski W., General Regression Neural Network (GRNN) in the Process of Tracking a Manoeuvring Target in ARPA Devices, *Proceedings of IRS 2005*, Berlin 2005.

Stateczny A., Kazimierski W.: Selection of GRNN Network Parameters for the Needs of State Vector Estimation of Manoeuvring Target in ARPA Devices, *SPIE Proceedings 2006*

Stateczny A., Kazimierski W.: The Process of Radar Tracking by Means of GRNN Artificial Neural Network with Dynamically Adapted Teaching Sequence Length in Algorithmic Depiction, *Proceedeings of 7th International Symposium of Navigation TransNav2007*, Gdynia 2007

Watson G. S.: Smooth regression analysis. *Sankhya Series A*, vol. 26, pp. 359–372, 1964.

24. The Modes of Radar Presentation of Situation in Inland Navigation

W. Galor
Maritime University of Szczecin, Poland

ABSTRACT: Navigation in inland waters has to meet the same requirements as those for pilot navigation in limited waters. This is due to the relation between the ship size and water area. Therefore, the requirements for position accuracy are much stricter than those in open sea offshore navigation. The radar is one of the devices for position determination. The most common mode of display showing the present situation on the screen is a panoramic display (bird's eye of view) similar to that of a navigational chart. The mostly radar panoramic presentation has a few disadvantages. In inland fairways and rivers this type of display does not fully satisfy navigators' requirements, because the water area perceived is small in relation to the surrounding land. The paper presents the idea of using of projection radar presentation based on orthogonal coordinates (bearing and distance).

1 INTRODUCTION

The main goal of navigation is to handle the ship in accordance with aim of their movement when required parameters of this process should be retained. The inland shipping requires the proper knowledge of navigators and adequate of navigation bridge equipment The process of ships movement in water area should be safely. Its estimation is executed by means of notions of safety navigation. It may be qualified (Galor W. 2009) as set of states of technical, organizational, operating and exploitation conditions and set of recommendations, rules and procedures, which when used and during leaderships of ship navigation minimize possibility of events, whose consequence may be loss of life or health, material losses in consequence of damages, or losses of ship, load, port structures or pollution of environment. Very often, the sea-river ships move on waterways (natural and artificial) inside of land for hundreds kilometres. The manoeuvring of ships on each water area is connected with the risk of accident, which is unwanted event in results of this can appear the losses. There is mainly caused by unwitting contact of ship's hull with other objects being on this water area. The safety of ship's movement can be identified as admissible risk, which in turn can be determined as combination of probability of accident and acceptable losses level (Galor w. 2009).

As a result, a navigational accident may occur as an unwanted event, ending in negative outcome, such as:

- loss of human life or health,
- loss or damage of the ship and cargo,
- environment pollution,
- damage of port's structure;
- loss of potential profits due to the port blockage or its parts,
- coast of salvage operation,
- other losses.

The inland waterways are restricted areas those where ship motion is limited by area and ships traffic parameters. Restricted areas can be said to have the following features:

- restriction of at least one of the three dimensions characterizing the distance from the ship to other objects (depth, width and length of the area),
- restricted ship manoeuvring,
- the ship has no choice of a waterway,
- necessity of complying with safety regulations set for local conditions and other regulations.

The floating craft (ship) during process of navigation has to implement the following safety shipping conditions:

- keeping the under keel clearance
- keeping the safety distance to navigational obstruction
- avoid of collision with other floating craft.

Thus the navigation on such waterways is different than on approaching waterways and coastal water areas. The realization of navigation on limited water areas is consisted on:

- planning of safety manoeuvre,

- ship's positioning with required accuracy on given area,
- steering of craft to obtain the safety planned of manoeuvre.

The leading of safely navigation requires first of all the high accuracy ship's positioning to avoid the contact with other ships and fixed objects. It can be natural objects (coast, water bottom) and artificial (water port structures-locks, bridges etc.) obstructers. Such kind of shipping is called as pilot's navigation. It necessitate the proper. The main elements of Integrated Bridge System are navigational systems-satellite, radar, electronic charts ECS/ECDIS. Navigation Integrated Bridge contains the devices of sips positioning (radio navigation systems including satellite) and presentation of situation (electronic charts system ECDIS, radar/ARPA) (Opracowanie...2009). A vessel's position in a restricted area should be considered as the position of its entire waterline area in the waterway. If the position of a manoeuvring ship is not known with required accuracy, there is a risk of navigational accident.

The distance between the hull and another object depends on the dimensions of required manoeuvring area within the waterway. For fairways the manoeuvring area is considered to be the width of vessel's swept path:

The navigational component of swept path width depends on:
- position determination accuracy,
- position determination frequency,
- methods of converting a position into the waterway coordinates.

The manoeuvring component depends on a number of factors. One of them is the time of the navigator's, i.e. pilot's or captain's response to observed movement off the fairway centre line, its analysis and giving a relevant command. The response time is affected by the same factors as those affecting the navigational component (mentioned above). The swept path reserve allows for hydrodynamic phenomena of bank effect or another object on vessel hull (mainly suction forces).

In the case of a system of continuous position determination, position determination accuracy is the basic element affecting the swept path width. That is why it is important to ensure that position determination is performed with appropriate (possibly highest) accuracy.

2 THE MODES OF RADAR PRESENTATION OF SITUATION

The radar is one of the basic devices which facilitate safe navigation in various conditions – both reduced and good visibility. The radar as a technical device significantly helps conducting a vessel by presenting a proper image of a situation around the vessel. The use of radio waves for presenting imaging objects enables a display of a situation that would be particularly difficult in poor visibility (fog, precipitation, night). In this way the radar facilitates steering a vessel in conditions in which human observation is much hampered, if not impossible (Fedorowski J. & Galor W. & Hajduk J. 1998). Nevertheless, radar observation also has some limitations resulting from the manner radar operates.

The use of radar for navigation can be said to have two basic goals:
- avoidance of collisions with stationary objects (natural objects such as the shore or bottom, and artificial objects such as port or other structures),
- avoidance of collisions with other vessels.

In both cases the operation of the radar can be divided into the following stages:
- detection of an object that results in a graphic presentation on the radar screen,
- object identification on the radar screen by the navigator,
- measurement of the detected and identified object (its position, movement parameters etc.).

The basic information for the navigator is presented on the display screen (Galor W. & Galor A. 2008). Presently, the display commonly used on board sea-going ships and other sailing craft is the type P display (called panoramic display). The display, showing a radio-located chart which illustrates the area surrounding the vessel, makes it possible to read out the range and direction (heading or bearing). Target echoes are displayed as spots displayed on the radar screen. Due to easy transformation of the polar coordinate system of the display into the Cartesian coordinate system of marine charts plus 'bird's eye view' imaging, the image interpretation is generally simple, except for a few particular situations. In spite of all the advantages of the panoramic display that make its use quite common, it should be noted that there are a number of shortcomings that limit substantially its range of applications. These are situations where navigation takes place in restricted areas, mainly rivers and channels or canals. When the range scale of observation is the same for the entire displayed area around the vessel, it often happens that the useless part of the screen (land beyond the shoreline) makes up 70% or more of the observed screen. Taking into account the width of a restricted area, the screen diameter (width) and the minimum operating range scale, it may turn out that using radar in such a situation is much more difficult.

The faults of that panoramic display may be largely eliminated by the method that presents the situation around the vessel as perspective display also known as type B (U.S. Radar...2006) perspective called sometimes also cineramic presentation (Brożyna J. 1984).

3 PANORAMIC PRESENTATION

Modern shipboard radars use mainly type P panoramic displays (PPI-Plan Position Indicator) for imaging information on the position of detected targets. Imaging on such a display resembles a chart and is 'drawn' in the polar coordinate system. Figure 1 presents a real image seen on the radar screen with the type P display recorded during a voyage of a vessel on the River Odra (Poland).

Each detected target is presented in position depended on its real distance (D) and bearing (NR). Formerly the radial –scan cathode tube ray was used. Than the co-ordinates of target position on screen were defined by distance and angle. Presently radars used screen where the picture is projected in raster – scan method. It means that position of targets is positioned in orthogonal co-ordinates XY. Thus the target dates achieved from radar sensor is transformed to these systems. The position of target A has a linear co-ordinate:

$$X_A = D \cdot \sin NR \tag{1}$$

$$Y_A = D \cdot \cos NR \tag{2}$$

where D= distance of target, NR=, bearing of target, X_A = linear horizontal co-ordinate, Y_A = linear vertical co-ordinate.

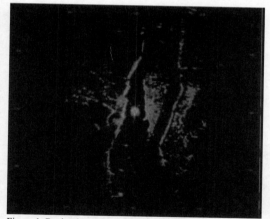

Figure 1. Real radar screen image with type P display

4 THE ASSESSMENT OF NAVIGATION BRIDGE OBSERVATION

In navigational practice there is often a need to compare a situation detected by the radar and displayed on the screen with a situation seen by the human eye. In this case the perceived image has geometry than that describing distance relation on a chart. How a picture observed by the human eye is created is shown on fig. 2 (Galor W. 2007), where A is an

apparent plane of the image, while the observer's eye can be seen on the left side of the diagram. The object O located at the distance A0 (section a can be neglected here as very small in comparison to A1) will be displayed as the point O1. The farther objects lie, the closer to each other and to the point H their images are seen, where the point H represents a point lying far away on the horizon. Therefore, comparison of the results of the eye perspective observation with a radar display image in the polar coordinate system calls for the transformation of the coordinate system. Then further actions can be performed, such as the object identification. From the navigator this transformation requires the capability of abstract thinking, which not always is possible, and is always an extra burden for navigator's mind, already loaded with a variety of duties. Besides, there is a risk that such transformation will be incorrect. The importance of this problem gets even greater if we realize that in practice navigators often has no time to analyze a situation with a pencil in their hand; then they make an overall estimation taking advantage of their experience and knowledge; on this basis navigational decisions are made. It is important that this experience is connected with the situation assessment in the display explained in figure 2. These factors are often a cause of frequent wrong interpretations of radar images; this, in turn, has resulted in a fact that in spite of placing fully operational radars on board ships the number of collisions has not been reduced.

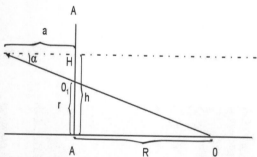

Figure 2. Visual assessment of a situation from the bridge.

5 PERSPECTIVE PRESENTATON

The perspective (called also as type B) shows radar information to the navigator with such geometry that he or she has been accustomed to since childhood as a regular observer, with a produced image similar to a TV or film picture. This imaging has a rectangular shape, where the horizontal axis is the bearing axis, while the vertical axis is the distance axis. Unlike a typical type P display used for other radiolocation purposes, the perspective display has a linear graduation on the horizontal axis, while on the vertical axis the graduation adequately intensifies towards the horizon Galor W. 2008). The target's co-ordinations (D, NR) are transformed on perspective presentation as:

$$X_A = D \cdot \sin NR \qquad (3)$$

$$Y_A = D \qquad (4)$$

In the panoramic display the range and direction have a rectangular representation. Such arrangement of the coordinates is often used in various areas in order to enhance distinguishing small angles, e.g. radar radiation characteristics at the horizontal cross-section. The method used for the presentation of a radar image allows to 'extending' the waterway, which will automatically reduce the useless part of the image on the screen. Figure 3 depicts a comparison of two screens with two different displays of types P and B. The examples show a situation for the bow sector covering -30^0, $+30^0$ areas. Naturally, the sector can be changed within the $0 - 90^0$ range. The 'O' marks heading line, 'A, B, C' indicate the different wide canals. It is evident that perspective presentation gives better relation of water area to total surface of radar screen.

Figure 4 shows a image of the same situation converted from P presentation into the type B display.

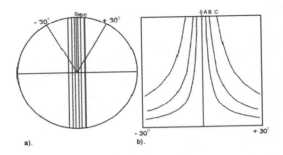

Figure 3. The radar screen: type P and B displays

The B display has much other beauty. One of them is easy and fast to detect of ships changing course from middle of the canal. In many cases the ships are going on axis of waterway. It is caused by hydraulic phenomena as suction by edge of canal (Projekt badawczy...2009). The B type display permits to fast positioning of ship in relation to the shore. Fig. 5 presents of such situation. The next beauty is fact, that the perspective presentation permits too much correction of angle's differentiation on the screen. It is correctly visible on fig.5 (right) for B display. The area near ship's position is stretched widely to P display. It highly permits for identification targets near the ship. Especially it respects the navigational marks (buoys, beacons), water structures (locks, bridges) and floating crafts (boats, yachts).

The above considerations on the method of presenting a situation of a vessel engaged in inland navigation justify the thesis that the method of vessel position determination (accuracy) and the mode of situation display are of major importance for navigators responsible for the process of navigation in inland waterways.

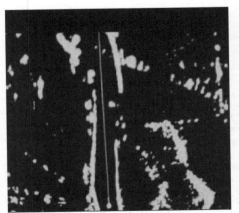
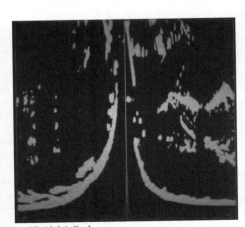

Figure 4. The real situation presented on screen radar - P (left) and B (right) displays

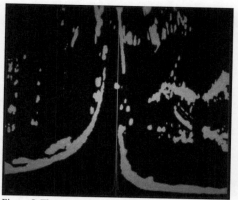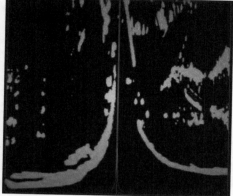

Figure 5. The presentation of ships course on right and left of axis waterway

6 SUMMMARY

The safety of navigation in restricted waters, particularly inland waterways, depends on the accuracy of determining the position of a vessel manoeuvring within such waters.

The radar is one of the devices for position determination. The most common mode of display showing the present situation on the screen is a panoramic display (bird's eye view) similar to that of a navigational chart.

This type of display does not fully satisfy navigators' requirements on fairways and rivers, because the water area perceived is small in relation to the surrounding land.

These shortcomings can be significantly reduced by applying a perspective display. Especially this kind of presentation of radar situation is useful in narrow water areas like canals, rivers in inland navigation.

The perspective presentation permits too much correction of angle's differentiation on the screen.

REFERENCES

Brożyna, J. 1984. Wskaźnik radarowy, *Patent No. 48610*, Polish Patent Office.

Fedorowski, J. & Galor, W.& Hajduk, J. 1988. Możliwości zastosowania wskaźnika radarowego typu B w żegludze pilotażowej i rzecznej. *Scientific Bulletin No 14 of Maritime University in Szczecin*, Szczecin.

Galor, W. 2007. Radar perspective display in inland navigation, *Polish Journal of Environmental Studies, Vol. 16, No 6B.*

Galor, W. & Galor, A. 2008. The using of radar by sea-river ships in inland navigation on lower part of Odra river, *Journal of KONES "Powertrain and Transport"*, Warsaw, pp 178-186.

Galor, W. 2008. Transformation of radar presentation for needs of sea-river navigation. *Advances in transport systems telematics,* Wydawnictwo Komunikacji i Łacznosci, Warszawa.

Galor, W. 2009a. The criterion of safety navigation assesment in sea-river shipping, *Marine Navigation and Safety of Sea Transportation,* publ. by CRC Press/Balkema, The Netherland.

Galor, W. 2009b. The 'certain value' approach to radar situation presentation in inland shipping. *Conference Book „Word Canal Conference 2009"*, Novy Sad, Serbia.

Projekt badawczy MNiSW N N508 3466 33. 2009. Opracowanie szczegółowych wytycznych do zaprojektowania zintegrowanego mostka nawigacyjnego jednostek w żegludze morsko- rzecznej. 2009. *Sprawozdanie,* Akademia Morska w Gdyni, Gdynia.

U. S. Radar Operational Characteristic of Radar Classified by Tactical Application. 2006. *Department of Navy,* Washington, USA.

Ship Handling and Ship Manoeuvering

25. Multirole Population of Automated Helmsmen in Neuroevolutionary Ship Handling

M. Łącki
Gdynia Maritime University, Gdynia, Poland

ABSTRACT: This paper presents the proposal of advanced intelligent system able to simulate and demonstrate learning behavior of helmsmen in ship maneuvering. Simulated helmsmen are treated as individuals in population, which through environmental sensing learn themselves to safely navigate on restricted waters. Individuals are being organized in groups specialized for particular task in ship maneuvering process. Neuroevolutionary algorithms, which develop artificial neural networks through evolutionary operations, are used in this system.

1 INTRODUCTION

One of the main tasks in Artificial Intelligence is to create the advanced systems that can effectively find satisfactory solution of given problem and improve it over time. Intelligent autonomous agents used in these systems can quickly adapt to current situation, i.e. change their behavior based on interactions with the environment (Fig. 1), become more efficient over time, and adapt to new situations as they occur.

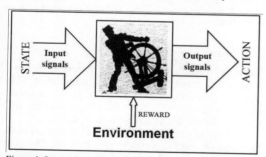

Figure 1. Interaction of helmsman with an environment.

Such abilities are very important for simulating helmsman behavior in ship maneuvering on restricted waters.

For simpler layouts learning process can be performed using classic approach, i.e. Temporal Difference Reinforcement Learning (Tesauro 1995, Kaelbling, Littman & Moore 1996) or Artificial Neural Networks with fixed structures (Braun & Weisbrod 1993). Dealing with high-dimensional spaces is a known challenge in Reinforcement Learning approach (Łącki 2007) which predicts the long-term reward for taking actions in different states (Sutton & Barto 1998).

Evolving neural networks with genetic algorithms has been highly effective in advanced tasks, particularly those with continuous hidden states (Kenneth & Miikkulainen 2005). Neuroevolution gives an advantage from evolving neural network topologies along with weights which can effectively store action values in machine learning tasks. The main idea of using evolutionary neural networks (ENN) in ship handling is based on evolving population of helmsmen.

The neural network is the helmsman's brain making him capable of choosing action regarding actual navigational situation of the vessel which is represented by input signals. These input signals are calculated and encoded from current situation of the environment.

In every time step the network calculates its output from signals received on the input layer. Output signal is then transformed to one of available actions influencing helmsman's environment. In this case the vessel on route within the restricted waters is part of the helmsman's environment. Main goal of the agents is to maximize their fitness values. These values are calculated from helmsmen behavior during simulation. The best-fitted individuals become parents for next generation.

2 NEUROEVOLUTION OF AUGMENTING TOPOLOGIES

Neuroevolution of Augmenting Topologies (NEAT) method is one of the Topology and Weight Evolving Artificial Neural Networks (TWEANN's) method (Kenneth & Miikkulainen 2002). In this method the whole population begins evolution with minimal networks structures and adds nodes and connections to them over generations, allowing complex problems to be solved gradually starting from simple ones.

The NEAT method consists of solutions to three main challenges in evolving neural network topology:

1 Begin with a minimal structure and add neurons and connections between them incrementally to discover most efficient solutions throughout evolution.
2 Cross over disparate topologies in a meaningful way by matching up genes with the same historical markings.
3 Separate each innovative individual into a different species to protect it disappearing from the population prematurely.

2.1 Genetic Encoding

Evolving structure requires a flexible genetic encoding. In order to allow structures to increase their complexity, their representations must be dynamic and expandable (Braun & Weisbrod 1993). Each genome in NEAT includes a number of inputs, neurons and outputs, as well as a list of connection genes, each of which refers to two nodes being connected (Fig. 2).

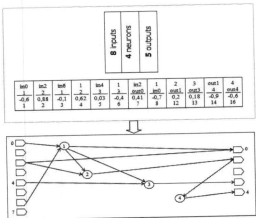

Figure 2. Genotype and phenotype of evolutionary neural network.

In this approach each connection gene specifies the output node, the input node, the weight of the connection, and an innovation number, which allows finding corresponding genes during crossover. Connection loopbacks are also allowed, as shown in figure 2.

2.2 Mutation

Mutation in evolutionary neural networks can change both connection weights and network structures (Fig. 3). Connection weights mutate as in any neuroevolutionary system, with each connection either perturbed or not.

Structural mutations, which form the basis of complexity, occur in two ways. Through mutation the genome can be expanded by adding genes or shrunk by removing them. In the add connection mutation, a single new connection gene is added connecting two previously unconnected nodes. In the add node mutation, the new node is placed, thus allowing to create new connections in future possible mutations.

Before connection genome mutation

After mutation

Figure 3. An example of weights and connection mutation in connections genome.

2.3 Crossover

Through innovation numbers, the system knows exactly what genes match up with each other. Unmatched genes are either disjoint or excess, depending on whether they occur within or outside the range of the other parent's innovation numbers.

In crossing over operation (Fig. 4), the genes with the same innovation numbers are lined up. The offspring is then formed in one of two ways: in uniform crossover, matching genes are randomly chosen for the offspring genome. In blended crossover, the connection weights of matching genes are averaged. These two types of crossover were found to be most effective in ENN in extensive testing compared to other crossover methods (Kenneth & Miikkulainen 2002).

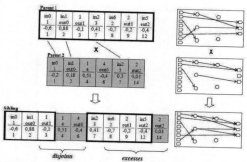

Figure 4. An example of crossover operation of equally fitted parents.

The disjoint and excess genes are inherited from the more fitted parent or from both parents, if they are equally fitted. Disabled genes (with zeroed weight) have a chance of being re-enabled during mutation, allowing networks to reuse older genes once again.

Evolutionary neural network can store history tracks of every gene in the population, allowing matching genes to be found and linked-up together even in different genome structures. Old behaviors encoded in the pre-existing network structure are not destroyed and remain qualitatively the same, while the new structure provides an opportunity to elaborate on these original behaviors.

In agent learning process, the genomes in NEAT will gradually get larger through mutation. Genomes of different sizes will sometimes result with different connections at the same positions. Any crossover operator must be able to recombine networks with differing topologies, to pass agents skills to next generations in meaningful way. Historical markings represented by innovation numbers allow NEAT to perform crossover without analyzing topologies. Genomes of different organizations and sizes stay compatible throughout evolution, and the variable-length genome problem is essentially avoided. This methodology allows NEAT to increase the complexity of the structure while different networks still remain compatible.

2.4 Speciation

Speciation of population can be seen as a result from the same process as adaptation (Beyer & Schwefel 2002), natural selection exerted by interaction among organisms, and between organisms and their environment (Spears 1995). Divergent adaptation of different populations would lead to speciation. Speciation of the population assures that individuals compete primarily within their own niches instead of competition within the whole population. In this way topological innovations of neural network are protected and have time to optimize their structure be-

fore they have to compete with other experienced agents in the population.

Generally, during species assigning process, as described in (Łącki 2009a), when a new agent appears in population, its genome must be assigned to one of the existing species. If this new agent is structurally too innovative comparing to any other individuals in whole population, the new species is created.

Compatibility of agent's genome g with particular species s is estimated accordingly to value of distance between two individuals. This distance is calculated with formula 1:

$$\delta = \frac{c_1 E}{N} + \frac{c_2 D}{N} + c_3 \overline{W} \qquad (1)$$

where: c1; c2; c3 - weight (importance) coefficients; E - number of excesses; D - number of disjoints; $\overline{W}$ - average weight differences of matching genes; N – the number of genes in the larger genome.

There must be estimated a compatibility threshold δ_t at the beginning of the simulation and if $\delta \le \delta_t$ then genome g is placed into this species. One can avoid the problem of choosing the best value of δ by making δ_t dynamic. The algorithm can raise δ_t if there are too many species in population, and lower δ_t if there are too few.

2.5 Fitness sharing

Fitness sharing occurs when organisms in the same species must compete with each other for life-sustaining resources of their niche. Thus, a species cannot afford to become too big even if many of its individuals perform well.

Therefore, any one species is unlikely to take over the entire population, which is crucial for speciated evolution to maintain topological diversity. The adjusted fitness f_i' for individual i is calculated according to its distance δ from every other individual j in the population:

$$f_i' = \frac{f_i}{\sum_{j=1}^{n} sh(\delta(i, j))} \qquad (2)$$

where: f_i – fitness value of individual i; sh - sharing function; n - number of individuals in whole population; - average weight differences of matching genes; $\delta(i,j)$ – distance between individuals i and j.

The sharing function sh is set to 0 when distance $\delta(i,j)$ is above the threshold δ_t; otherwise, $sh(\delta(i,j))$ is set to 1 (Spears 1995). Thus, sum of sh calculates the number of organisms in the same species as individual i. This reduction is natural since species are already clustered by compatibility using the threshold δ_t. A potentially different number of offspring is assigned to every species. This number is proportional to the calculated sum of adjusted fitness values f_i of its members.

In the first step of species reproduction process the system eliminates the lowest performing members from the population. In the next step the entire population is replaced by the offspring of the remaining organisms in each species (Fig. 5).

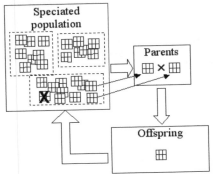

Figure 5. Evolution within one species in speciated population.

The other selection methods in speciated population are also considered in future research, i.e. island selection or elite selection of best fitted individuals of every species with particular task.

The final effect of speciating the population is that structural innovations are protected.

3 MULTIROLE SHIP HANDLING WITH EVOLUTIONARY NEURAL NETWORK

The main goal of authors work is to make a system able to simulate a set of navigational situations of ship maneuvering through a restricted coastal area. This goal may be achieved with Evolutionary Neural Networks (ENNs).

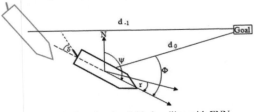

Figure 6. Sample data signals of ship handling with ENN.

Navigational situation of a moving vessel can be described in many ways. Most important is to define proper state vector from abundant range of data signals (Fig. 6) and arbitrary determine fitness function values received by the agent. Fitness calculation is of primary meaning when determining the quality of each agent. Subsequently it defines helmsman's ability to avoid obstacles while sailing toward designated goal.

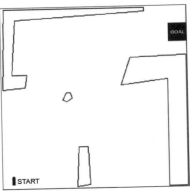

Figure 7. Model of simulated simplified coastal environment.

In the simplified simulation model there are no moving vessels in the area (Fig. 7) (Łącki 2009b).

Helmsman observes current situation which is mapped into input signals for his neural network. In general considered input signals indicating (i.e.):
- The ship is on the collision course with an obstacle,
- Ships course over ground,
- Ships angular velocity,
- Distance to collision,
- The ship is approaching destination,
- Ships angle to destination,
- The ship is heading out of the area,
- Danger has increased,
- Danger has decreased,
- Ship is heading on goal.

All the input signals are encoded binary (Łącki 2010a). Neural network output value is the rudder angle. It is crucial for effectiveness of simulation to determine the number of neural network outputs.

More outputs mean more calculations but on the other hand better accuracy and fidelity of designed environment. Additionally too many outputs increase complexity of the learning process, thus making an agent unable to quickly adapt to new situations. This accuracy vs. performance dilemmas were examined extensively in previous works (Łącki 2008-2010).

The fitness value of an individual is calculated from arbitrary set action values, i.e.:
- -1 for increase of the distance to goal in every time step,
- -10 when ship is on the collision course (with an obstacle or shallow waters),
- +10 when she's heading to goal without any obstacles on course,
- -100 when she hits an obstacle or run aground,
- +100 when ship reaches a goal,
- -100 when she departs from the area in any other way, etc.

In the simplified simulation model speed of the ship was constant. In the advanced model, in which one considers a set of possible task, there must be possibility for the agent to adjust speed of the vessel. Situation evaluation can be treated as multi-criteria problem which calculates a danger of getting stranded on shallow water, encountering a vessel with dangerous cargo, getting to close to shore, etc. It can be estimated with available optimization algorithms (Filipowicz, Łącki & Szłapczyńska 2005) with function of ship's position, course and angular velocity, information gained from other vessels (if considered in the model) and coastal operators.

3.1 Multirole ship handling system

Situation Determining Unit (SDU) is an intelligent module responsible for grouped helmsmen management (Fig. 8).

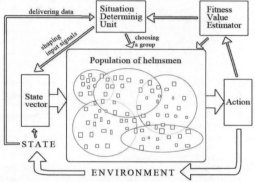

Figure 8. Complementary multirole neuroevolutionary ship handling system.

This unit constantly receives information from state of the environment, processes it, determines actual navigational situation and regarding it chooses the best fitted group of helmsmen for this task. SDU is designed with neuroevolutionary system. The best performing neural network is the one making decisions. Its performance is determined regarding fitness values reinforcing the neural network by Fitness Value Estimator. With elite selection method this neural network participates in creation of new generation of SDU's.

State vector is dynamic in that its signals depend on current navigational situation chosen by SDU. Set of available actions is determined in the same way. An example of input and output signals for general navigation task is presented on figure 9.

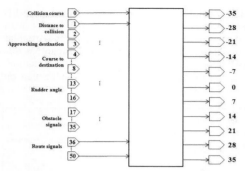

Figure 9. Input and output signals in general navigation task.

List of considered possible tasks:
- General navigation – during this task SDU observes possible collision warnings. All the vessels are treated as points. Action: coarse rudder angle.
- Collision avoidance – The vessels involved in collision situation are treated as 2D objects. Action: precise rudder angle, propeller thrust control,
- Turning – Action: rudder deflection, propeller thrust and bow/stern thrusters' control,
- Mooring – Action: propeller thrust and bow/stern thrusters' control, rudder deflection.

Two different multirole models of restricted waters environment can be taken into consideration in the advanced system:
- Single vessel multirole population simulation,
- Multi-task multi-agent simulation.

In the first model the main goal is to train population of helmsmen to safely handle a particular model of a ship in specified single multirole task (i.e. safe passage trough congested water channel, approach to dock and mooring). In this case agents compete simultaneously with each other in one population (may be grouped into species) to handle a ship as best as possible. Any other vessels in the area are treated as stationary or moving objects on reasonably predictable routes (Fig 10).

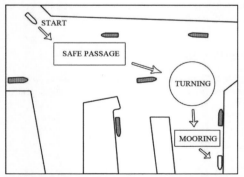

Figure 10. Example of an environment with simultaneous multirole population of helmsman.

The second model consist a small number of different vessels in the same restricted waters with helmsmen as the best trained agents for particular ships (Fig. 11). In this case every helmsman has different goal and different environmental situation, depending on actions taken by other helmsmen on the other vessels.

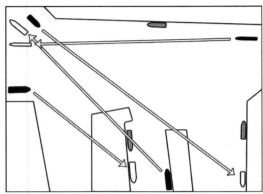

Figure 11. Example of multi-task multi-agent environment. There are four agents allocated on four vessels. Starting points of the vessels are indicated with black outlines while goals for them are marked with white ones.

4 REMARKS

Neuroevolution approach to intelligent agents training tasks can effectively improve learning process of simulated helmsman behavior in ship handling (Łącki 2008). Neural networks based on NEAT increase complexity of considered model of ship maneuvering in restricted waters.

Implementation of multirole division of helmsmen population in neuroevolutionary system allows simulating complex agents' behavior in the environments with much larger state space than it was possible in a classic state machine learning algorithms (Łącki 2007). In this system it is also very important to change parameters of genetic operations dynamically as well as the input signals vector and set of available actions.

REFERENCES

Beyer, H.-G. & Schwefel, P. H. 2002. *Evolution strategies – A comprehensive introduction*. Natural Computing, 1(1):3–52.

Braun, H. & Weisbrod, J. 1993. *Evolving feedforward neural networks*. Proceedings of ANNGA93, International Conference on Artificial Neural Networks and Genetic Algorithms. Berlin: Springer.

Chu T. C., Lin Y. C. 2003. *A Fuzzy TOPSIS Method for Robot Selection*, the International Journal of Advanced Manufacturing Technology: 284-290,

Filipowicz, W., Łącki, M. & Szłapczyńska, J. 2005, *Multicriteria decision support for vessels routing,* Proceedings of ESREL'05 Conference.

Kaelbling, L. P., Littman & Moore. 1996. *Reinforcement Learning: A Survey.*

Kenneth, O.S., & Miikkulainen, R. 2002. *Efficient Evolution of Neural Network Topologies*, Proceedings of the 2002 Congress on Evolutionary Computation, Piscataway.

Kenneth, O.S. & Miikkulainen R. 2005. *Real-Time Neuroevolution in the NERO Video Game*, Proceedings of the IEEE 2005 Symposium on Computational Intelligence and Games, Piscataway.

Łącki, M. 2007 *Machine Learning Algorithms in Decision Making Support in Ship Handling*, Proceedings of TST Conference, Katowice-Ustroń, WKŁ.

Łącki, M. 2008, *Neuroevolutionary approach towards ship handling*, Proceedings of TST Conference, Katowice-Ustroń, WKŁ.

Łącki, M. 2009a, *Speciation of population in neuroevolutionary ship handling*, Marine Navigation and Safety of Sea Transportation, CRC Press/Balkema, Taylor & Francis Group, Boca Raton – London - New York - Leiden, p. 541-545.

Łącki, M. 2009b, *Ewolucyjne sieci NEAT w sterowaniu statkiem*, Inżynieria Wiedzy i Systemy Ekspertowe, Akademicka Oficyna Wydawnicza EXIT, Warszawa, p. 535-544.

Łącki, M. 2010a, *Wyznaczanie punktów trasy w neuroewolucyjnym sterowaniu statkiem*. Logistyka, No 6.

Łącki, M. 2010b, *Model środowiska wieloagentowego w neuroewolucyjnym sterowaniu statkiem*. Zeszyty Naukowe Akademii Morskiej w Gdyni, No 67, p. 31-37.

Spears, W. 1995. *Speciation using tag bits*. Handbook of Evolutionary Computation. IOP Publishing Ltd. and Oxford University Press.

Stanley, K. O. & Miikkulainen, R. 2002. *Efficient reinforcement learning through evolving neural network topologies*. Proceedings of the Genetic and Evolutionary Computation. Conference (GECCO-2002). San Francisco, CA: Morgan Kaufmann.

Stanley, K. O. & Miikkulainen, R. 2005. *Real-Time Neuroevolution in the NERO Video Game*, Proceedings of the IEEE 2005 Symposium on Computational Intelligence and Games, Piscataway.

Sutton, R. 1996. *Generalization in Reinforcement Learning: Successful Examples Using Sparse Coarse Coding*. Touretzky, D., Mozer, M., & Hasselmo, M. (Eds.), Neural Information Processing Systems 8.

Sutton, R. & Barto, A. 1998. *Reinforcement Learning: An Introduction.*

Tesauro, G. 1995. *Temporal Difference Learning and TD-Gammon*, Communications of the Association for Computing Machinery, vol. 38, No. 3.

26. Ship's Turning in the Navigational Practice

J. Kornacki
Szczecin Maritime University, Szczecin, Poland

ABSTRACT: The turning-basin is a special part of the port waters area. During every port's call ship need to be turned. The paper presents the analysis of the ship's turning in the navigational practice. The paper is based on the practical experience.

1 INTRODUCTION

The manoeuvre of ships turning is one of the most frequently executed port manoeuvres. It is executed every time during the ships port call. Simultaneously, it is comparatively little examined manoeuvre.

Place, where the ship's turning manoeuvres are executed, is called the turning basin. It can be understood as a two different meanings. First as the manoeuvring area appointed by the ships, place needed for a ship to execute turning. Second meaning is the hydro-technical building artificial or natural with suitable horizontal and vertical dimensions, where the considerable alterations of the course of the ship are executed. Shortly, it is the hydrotechnical construction, part of the port waters. Certainly due to safety, the turning basin as the hydro technical building always has to be larger in all dimensions than the manoeuvre area to avoid the collision with bottom or bank (Kornacki & Galor 2007).

Figure 1. The ships turning manoeuvre with using of two tug-boats.

The main parameter describing the turning basin is a size of the turning basin and size of manoeuvring area needed to execute manoeuvre. The influences on the size of turning basin during the manoeuvre have the large quantity of factors.

The ship's turning manoeuvres in a practice are „in the place". This should be understand as the changing (alternation) of the ship's course whose longitudinal velocities, during the manoeuvre, are close to zero (Kornacki 2007).

Turning the ship over the place is done on the turning basin as a result of the planned tactics of manoeuvring and can be done by the ships itself or in co-operation with tugs or use of anchors or lines (springs).

To be able to describe the practical realization of the manoeuvre in the quantitative way, the factors and the ways of the assessment of the manoeuvre will be introduced. The attempt at the exact assessment of manoeuvres will not be undertaken. The "operator" executing manoeuvre puts the actions to continuous evaluation of the correctness and the advisability of the next steps during of his work.

The acquaintance of the phenomenon of the instantaneous pivot point is important during executing the ship's turning manoeuvres. The existence of this phenomenon in ship manoeuvring is known to navigators. The recognizing of the instantaneous pivot point's location on a ship during ship's turning manoeuvres is causing some problems. It is important to be well-informed about location.

In the last part, the assay of the presentation of the practical view of the realization of the manoeuvres of the ship's turning will be undertaken. The attempt at presenting will be undertaken as the decision process proceeding and as the decision transfers on the realization of the manoeuvre.

2 THE ASSESSMENT OF THE SHIPS TURNING MANOEUVRE

Wanting to evaluate the realization of the manoeuvres of the turning, the methodologies of such assessment should be known.

The opinion, that the manoeuvre has to be executed effectively and safely, is repeated during the various conferences and discussion on the subject of the skill on shiphandling generally and shiphandling on the simulators. One can show on three elements which are very essential for the statement the correctness's of the realization of the manoeuvre on the basis of the convention STCW'78 with later corrections and experience from led trainings on the shiphandling simulator (Artyszuk 2002). The elements of assessment of such manoeuvres should be the statement that the manoeuvre was executed:
- safely
- effectively
- according to the principles of the manoeuvring practice

The assessment basis on this opinion can be proposed as the sum of these elements (Kornacki & Kozioł 2006):

$$O \approx B + E + Z \qquad (1)$$

where:
O - the assessment of the manoeuvres,
B - safety of the manoeuvres,
E - effectiveness of the manoeuvres,
Z - compatibility with the principles of the manoeuvring practice.

One can propose the following division the criteria, which will have the influence on the assessment of the realization of the given element, and in the effect of the whole manoeuvre.

The element of the safety of the manoeuvre is estimated on the basis of following criteria:
- the size of the area of manoeuvring the ship during the of executing the manoeuvre,
- appearing the contact with different individual or navigational obstacle during the of executing the manoeuvre,
- the energy of the collision with bottom, underwater slope, bank (it is sometimes admissible such contact depending on the accepted tactics of the manoeuvre).

Criteria such as: the size of the manoeuvring area and the energy of impact (collision) they are applied in the investigations of Marine Traffic Engineering to the assessment of safety during port manoeuvres at present. Can, therefore also use these criteria as the criteria of the assessment of the skill of manoeuvring (Guziewicz & Ślączka 1997).

The element of the realization of the manoeuvre according to the principles of the manoeuvring practice should comply following criteria:
- the using of the rudder,
- the using of propulsion,
- the using of maximum sets of propulsion,
- the time of the realization of the manoeuvre,
- the using of the tugboats,
- the using of the anchors,
- the using of the lines.

The element of the effectiveness of the manoeuvre is estimated in the moment of the end of manoeuvre on the basis of following criteria:
- final ship's position on the end of manoeuvre,
- ship's heading on the end of manoeuvres,
- ship's speed and yaw velocity on the end of manoeuvres (if it is acceptable for continue with next manoeuvres),

All of these elements have to be acceptable for next manoeuvres. Turning manoeuvres are always between the others.

The element of effectiveness is the assessment of the last stage of the manoeuvres, when the ship is in the peaceable area with the aim of the given manoeuvre.

The very reliable criterion for the assessment the skill is the quantity of given commands in relation to the last element of the realization of the manoeuvre. The number of given commands in the reference to rudder, propulsion, using of the anchors or tugboats are the objective coefficient acquired the practice of ship's manoeuvring. Mostly, the persons manoeuvring worst are giving too much commands and so often and the manoeuvres become less controlled. The problem of the good manoeuvring practice as the good seamanship is not simple. The giving of emotional and incessant commands, often before the reaching the result of previous is the bad practice. The regard of the parameter relating pronouncement of maximum sets is important because of the fact of the influence of propeller's streams on the bottom of the reservoir and quay. However, leaving of the reserve of the power during the executing manoeuvres is the good practice in reference to the principles of manoeuvring, which can turn out indispensable in emergency situations or sudden influence unforeseen external conditions.

The assessment of individual elements is executed in the moment of the end of the manoeuvre.

Taking under considering above mentioned, it should be added that the assessment of these elements has to contain their weights.

Analysing the assessment, it can be introduce as follows (Kornacki & Kozioł 2006):

$$O = w_B * \sum_{i=1} b_i + w_E * \sum_{j=1} e_j + w_Z * \sum_{l=1} z_l \qquad (2)$$

where:

w_B - the safety elements' weight,

w_E - the effectiveness elements' weight,

w_Z - the weight of compatibility with the principles of the manoeuvring practice,

b_i - the criterions of safety,

e_j - the criterions of effectiveness,

z_l - the criterion of compatibility with the principles of the manoeuvring practice.

Economical is the additional element which in the practice will have the influence on the assessment of every manoeuvre. Also, the manoeuvre has to be optimum in relation to economic. For sure, the glaring abuse of the means will be disqualifying.

Summing up, the problem of the assessment of the skill of manoeuvring is complex. Executing the manoeuvre of the turning, the previously suitable tactics is accepted. During the executing of the manoeuvre of the turning and on the end, the manoeuvre is the subject to the assessment by performers on every stage to improve possible mistakes.

3 PIVOT POINT IN SHIP MANOEUVRING

The acquaintance of the phenomenon of the instantaneous pivot point is important during executing the manoeuvres of the ship's turning. This phenomenon existence in ship manoeuvring is well known to navigators. The recognizing of location of the instantaneous pivot point on a ship during various manoeuvres is causing problems for them. For a ship's operator is easier to control just a single motion and practical role of instantaneous pivot point increase. The navigators are familiar with apply the pivot point principles while making various kinds of manoeuvres, either at high or at low forward speed.

This is the essence of the kinematics of the turn of the ship that certain lateral velocity of the ship's centre of gravity always accompanies the angular velocity. It is directed in the external board in the relation to the centre of the circulation. During the turn, local lateral velocities on the stern and the bow of the ship are different from this existing in the centre of gravity. The lateral speed is larger on the stern, on the bow meanwhile smaller. It results from this, that on the stern the local lateral velocity coming from the rotational movement adds to the fine lateral movement, while it is on the bow inversely. The point in which the local lateral velocities reduce to zero is the instantaneous pivot point. The complex movement of the ship, being in the generality connection of three movements i.e. longitudinal, lateral and rotational movements in relation to the centre of gravity can replace with two movements consisting

from the longitudinal and rotational movement in relation to the instantaneous pivot point.

At high speed applying the stern rudder will involve the rudder lateral force as developed on a constant arm. The ship's lateral motion strongly correlated with simultaneous turning is generated, consequently the instantaneous pivot point, in certain time, lies somewhere close to the bow and usually changes but in a rather narrow limit. Practically, the pivot point can be considered steady location.

3.1 *The position of instantaneous pivot point*

It is well known from rigid body mechanics that instantaneous arbitrary ship planar motions can be uniquely decomposed into a combination of translational motion of the point of origin and angular motion of the whole body around this origin. Both motions are defined by the linear velocity vector of the origin and the angular velocity vector. Such an instantaneous state of motion is equivalent to a sole angular motion around the instantaneous pivot point. The pivot point is a point on the body or can be even outside the body physical extents, in which the linear velocity disappears. The pivot point position instantaneously changes because of the ships linear and angular velocities are time dependent.

The position of instantaneous pivot point reference to the centre of gravity or amidships is qualified exact kinematic dependence (Artyszuk 2009):

$$x_{PP} = \frac{-v_y}{\omega_z} \text{ and } y_{PP} = \frac{v_x}{\omega_z} \qquad (3)$$

or non-dimensionally - divided by ship's length L:

$$x'_{PP} = \frac{-v_y}{\omega_z L} \text{ and } y'_{PP} = \frac{v_x}{\omega_z L} \qquad (4)$$

where:

x_{PP} - the longitudinal coordinate of pivot point,

x'_{PP} - the longitudinal non-dimensional coordinate of pivot point,

y_{PP} - the lateral coordinate of pivot point,

y'_{PP} - the lateral non-dimensional coordinate of pivot point,

v_x - the longitudinal velocity,

v_y - the lateral velocity,

ω_z - the angular velocity,

L - the length of the ship.

The dimensionless values relating to any linear dimension of the ship is much more useful also in the shiphandling practice. The tips are universal for any ship size and ship visual positioning against external objects is also easier in relative units.

The instantaneous pivot point lies mostly opposite the side of the place of applying steering forces. During manoeuvres with the stern rudder or as a result of the lateral effect of the propeller during active stopping the ship, it is placed to the bow from amidships. The position of instantaneous pivot point is more-less steady for the given individual ship, although it is able to change in certain borders. The position of the instantaneous pivot point is comprises in range of $0.3 \div 0.5\ L$ counted from amidships (or $0 \div 0.2\ L$ counted from bow). Often, the literature tells about average values of this position for the ships in range of $0.25 \div 0.3\ L$ counted from a bow, it should be treat rather for orientation and the current values should be accepted according to the parameters of the circulation settled e.g. (Artyszuk 2009).

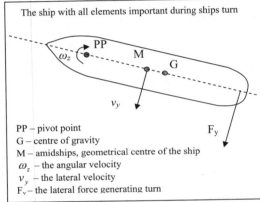

The ship with all elements important during ships turn

PP – pivot point
G – centre of gravity
M – amidships, geometrical centre of the ship
ω_z – the angular velocity
v_y – the lateral velocity
F_v – the lateral force generating turn

Figure 2. The ship with lateral force generating turn.

Often, it is very convenient for a short prediction period, as dominating in shiphandling practice, to disregard the lateral non-dimensional coordinate of pivot point and, instead of, to rotate a ship around the pivot point projected on the ship's centre line together with its simultaneous translation in the direction of heading according to the forward speed.

It is two other aspects of position of instantaneous pivot point. First is a position of it during turning of the ship in spot. Second is position of it during astern movement.

3.2 The position of instantaneous pivot point during ship turning in spot

Typical situation of ship turning at spot is while the longitudinal velocity is nearly zero, the instantaneous pivot point is located on the ship's centre line and the constant yaw velocity is requested. This kind of manoeuvres happens in turning basins.

To realize turning a ship at spot, the following steady phase motion equations shall be (Artyszuk 2010):

$$
\begin{cases}
(m + m_{11})\dfrac{dv_x}{dt} = F_x + (m + c_m m_{22})v_y \omega_z = 0 \\[2mm]
(m + m_{22})\dfrac{dv_y}{dt} = F_y + F_{yH} = 0 \\[2mm]
(J_z + m_{66})\dfrac{d\omega_z}{dt} = M_z + M_{zH} = 0
\end{cases}
\tag{5}
$$

where:
m_{11}, m_{22}, m_{66} - virtual masses [kg], [kg], [kg m^2],
v_x, v_y, ω_z - surge, sway and yaw velocity [m/s], [m/s], [1/s],
F_x, F_y, M_z - external total surge, sway forces and yaw moment [N], [N], [Nm],
F_{yH}, M_{zH} - hull sway forces and hull yaw moment [N], [Nm],

In case of application of two independent parallel forces, like using of two (or more) tugboats, such forces and moments can be defined by:

$$
\begin{cases}
F_y = F_{y1} + F_{y2} \\[2mm]
M_z = F_{y1} x_{F1} + F_{y2} x_{F2}
\end{cases}
\tag{6}
$$

This should solve the problem of discrete attachment points for tugs.

The turning with particular value of position of instantaneous pivot point can be completed with a single external force positioned between the ship's bow and stern (Artyszuk 2010):

$$
x'_F = \frac{M_z}{L F_y}
\tag{7}
$$

The use of tugs (one or more) for large vessels as to generate the required external force is limiting the arms of external forces like technically possible contact points around the ship's hull for tugboats, like in pushing mode, where tugs are allowed in specially marked places contact with the hull or in pulling mode, where tug operation is restrained by the arrangement of ship's fairleads.

3.3 The position of instantaneous pivot point during astern movement

The ship's propeller working astern generates a side effect called also "wheel effect". This phenomenon is mainly joined with the inflow the reverted propeller race to the ship's stern and pressure characteristic of water medium generating lateral trust. Propeller generate twisted slipstream attacking various sections of ship's stern. As long as ship's positive speed is remaining the pivot point stay forward of amidships. It is proceeding independently from the astern throttle. While a ship starts a sternway the pivot point location changes astern of amidships. The pivot point location sign is changing from a positive to a negative one. This behaviour is not fully explained. Because of the propeller lateral force continues to keep its direction and generate the accompanied ship's lateral velocity, the ship's centrifugal force seems to be primarily reason of this behaviour. Anyway, this shifting of the pivot point from forward to stern because of continued propeller astern trust together with astern velocity and lateral velocity base on observation. For sure, the suitably large external force on the ship's hull will cause new moving the pivot point in dependence from the point of applying and size of the applied force.

4 THE TECHNIQUE OF MANOEUVRING

Turning manoeuvres of the ship could be systematized in relation to the aim of executed manoeuvres (which is not relevant), the conditions for manoeuvre (hydro-meteorological conditions, bathymetric conditions, legal aspect) that affect the realization of the manoeuvre, and most important, way of execution of manoeuvre, which can be further divided into independent manoeuvres and manoeuvres with tugboats assists. During the ship's turning manoeuvres is an important use of available propellers, anchors, ropes or tugs to manoeuvre was executed an manoeuvre area as small as possible while the use of the propellers, tugs, etc. was the most optimal in different ways.

In the practice, time of realization of such manoeuvre amounts from a few to more than ten minutes in dependence from accessible propellers and propulsions, the possibility of the use of the accepted tactics of manoeuvring, the ship's condition, hydro-metrological conditions and bathymetric conditions.

4.1 The independent ship's turning maneuvers

Independent manoeuvres can be executed using:
- only screw propeller and stern rudder,
- the bow thrusters and screw propeller and stern rudder,
- the bow and stern thrusters and screw propeller and stern rudder,
- anchor and the bow thrusters (if available) and screw propeller and stern rudder,
- lines (springs) and the bow thrusters (if available) and screw propeller and stern rudder.

The instantaneous pivot point moves from the extreme position in the bow area (during ahead propeller's mode) to the position in the stern area (while astern movement) during independent executing the turning manoeuvre (Rowe 1996). This changeability of the position of the instantaneous pivot point during the manoeuvre is effective on larger manoeuvring area of the ship's turning. The occurrence of the longitudinal velocity alternating forward and astern is additional factor influencing the increase of the manoeuvring area.

The use additionally anchors or lines (springs) does not cause considerable decrease of the sizes of the manoeuvring area. What is the aim of their applying? Simply, it would not be able to execute successfully such manoeuvre in many situations. It seems that use of anchors or lines should influence considerably of the size of the manoeuvring area. These elements, in practice, influence directly of the turning ability and their influence on the size of the manoeuvring area are not great and grow up together with the growth of the influence of external (hydrometeorological) factors what simulating investigations confirm. The quantity of applied commands is significantly decreasing facilitating the realization and in others time of realization of such manoeuvre is significantly decreasing. Often, it is the large meaning on current narrow waters.

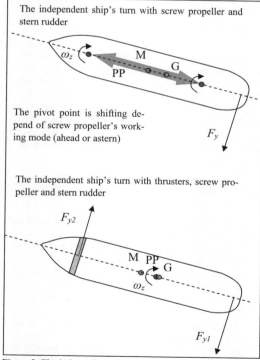

Figure 3. The independent ship's turning.

Thrusters are significant support for the ship's turning manoeuvres. They give the possibility the use constant forces apply in extreme points and affecting in opposite directions. The instantaneous pivot point does not change the position practically, and the yaw moment increase significantly. The additional using of thrusters helps significantly decrease the time, quantity of given commands and the sizes of the manoeuvring area.

4.2 *The ship's turning maneuvers with tugboats assists*

Tugboats, similarly as thrusters, are a significant support for the manoeuvres of the turning of the ship.

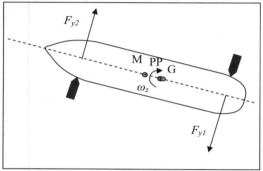

Figure 4. The ship's turning with tugboats assists.

They can work with long towing-hawsers, short towing-hawsers and as pushers. They give, in the principle, such advantages and the effects of working as in case of use of thrusters. The possibility of use of various directions of working tugboats is the additional advantage. The only deficiency can be the sizes of the manoeuvring area caused addition to the manoeuvring area of ship, the manoeuvring area of tugboats sometimes.

4.3 *The ship's turning maneuvers with other advanced propellers solutions*

Turning manoeuvres of ships with other advanced propellers solution signify:
- turning manoeuvres with twin screw propellers (with or without thrusters accompany),
- turning manoeuvres with advanced rudders,
- turning manoeuvres with azimuth drive,
- turning manoeuvres with Voiht-Schneider drive.

This advanced propellers solutions are applied mostly on smaller ships. This is the result of the aim of exploitation and costs. They are combining and increasing advantages of all independent ship's turning and ship's turning with tugboats assists, simultaneously decreasing deficiencies. The forces and yaw moments are enlarged by multiplying of different propellers and application more effective rudders.

This kind of propulsions brings great decreasing of time consumption in ship's turning. Short time of manoeuvring transfers to safety of manoeuvring. It is minimizing the ship's manoeuvring area.

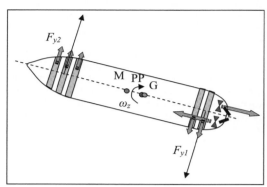

Figure 5. The ship's turning with advanced propellers solution (twin screw propellers, five thrusters).

Appling of advanced propellers solutions brings some danger of human error in service. Some time, there are a lot of different elements that could be regulated. On some of them, to avoid any errors, advanced steering modules with consoles are installed. Then operator needs to declare direction and force of working of the propulsion.

5 CONCLUSIONS

The navigator manoeuvring the ship on the turning basin has concretely presented the problem which has to execute with the success. He has the ship with definite technical solutions and the accessible waters with known conditions where the task has to be made.

While executing the manoeuvre the process which he operates still assesses.

Many factors which concentrate in three groups consist on the assessment of the ship's turning manoeuvre. All three elements have to be fulfilled in the minimum degree to recognise the ship's turning manoeuvre correctly executed.

The variety of solutions carries with her the plurality of the possibility of their using.

The direct influence on the realization of the ship's turning manoeuvre has position of the instantaneous pivot point. The instantaneous pivot point depends of the way how the operator steers propulsion and the accessible solutions of drives.

The techniques of manoeuvring influence the position of the instantaneous pivot point, longitudinal velocity and the time of the realization of the manoeuvre. These factors influence the size of the manoeuvring area. The manoeuvring area influences the safety of the executed manoeuvre. The operator can suitably influence the safe realization of the manoeuvre assessing the development of the situation and suitable action.

In practice, executing the manoeuvre of the ship's turning all possible resources are used. This secures the development of the safety and the certainty of the realization of the manoeuvre. The plurality of steerable resources brings with them the development of the danger of the pronouncement of the human error.

The economics influences the manoeuvre of the turning significantly. For ships execute this manoeuvre seldom using of the tugboats assist is more justified. Better propulsion and steering equipment of the ship is more appropriate for ships executing this manoeuvre often.

REFERENCES

Artyszuk, J. 2002. Practical Evaluation of Shiphandling Skills. TRANSCOMP '2002, Prace Naukowe 'Transport' nr 1(15), Politechnika Radomska.

Artyszuk, J. 2009. Revised Guidelines on Instantaneous Pivot Point of a Berthing Ship. Molme: Conference on Marin Traffic Engineering, Maritime University of Szczecin.

Artyszuk, J. 2010. Pivot point in ship manoeuvring. Assessment. Scientific Journals (92) Maritime University of Szczecin.

Gucma, S. & Jagniszczak, I. 2006. Nawigacja dla kapitanów. Gdańsk: Fundacja Promocji Przemysłu Okrętowego I Gospodarki Morskiej.

Guziewicz, J. & Ślączka, W. 1997. Szczecin: The methods of assigning ship's manoeuvring area applied in simulation research, International Scientific and Technical Conference on Sea Traffic Engineering, Maritime University of Szczecin.

Kornacki, J. & Kozioł, W. 2006. Szczecin: Selected Problems of Shiphandling Skills Assessment. Scientific Journals (80) Maritime University of Szczecin.

Kornacki, J. & Galor, W. 2007. Gdynia: Analysis of ships turn manoeuvres in port water area. TransNav'07.

Kornacki J. 2007. The analysis of the methods of ship's turning-basins designing. Szczecin: XII International Scientific and Technical Conference on Marin Traffic Engineering, Maritime University of Szczecin.

Rowe, R.W. 1996. London: The Shiphandler's Guide. The Nautical Institute.

Vasco Costa, F. 1968. Berthing Manoeuvres of Large Ships. The Dock and Harbour Authority (48).

Search and Rescue

27. New Proposal for Search and Rescue in the Sea

I. Padrón Armas
c/ Guaydil No 68 Tamarco, Tegueste, Tenerife, Espana

D. Avila Prats, E. Melón Rodríguez, I. Franquis Vera & J. Á. Rodríguez Hernández
Dpto. CC y TT de la Navegación (ULL) C/ Padrón Albornoz s/n, S/C

ABSTRACT: Currently the security of a ship would be reduced if among its systems don't have some able to indicate its position, providing the necessary data to try that the Rescue System begin a rescue in the event of emergency or catastrophe. Formerly the localization of a ship was based on warnings transmitted by the sailors or automatic systems. The efficiency of these systems was demonstrated in the rescue services and salvage of human lives obtained along more than a century. Since final of the last century the marine transport introduced a new aid system, the Global Maritime Distress Safety System (GMDSS), based in the automatic alarms through communications for satellites and the radiobeacons use that transmit data that help to the localization of a ship.
In this work is sought to characterize some of the systems of satellites more diffused for the localization of aid signs (INMARSAT, COSPAS-SARSAT), adding the project IRIDIUM for improvement of the search and rescue, by means of the surveillance from satellites in orbit. A comparison will be established among these systems, taking parameters like: global covering, emission of false alarms, portability and economic cost, with the objective of determining the most effective system in case of catastrophes.

1 INTRODUCTION

The beginning of the marine communications at distance born with the use of the telegraphy without threads, the localization of a ship was based on warnings transmitted by the sailors or automatic machines. The effectiveness of these systems was demonstrated in the rescue services and salvage of human lives, during more than a century. Currently the security of a ship would be reduced, if enters its systems don't have some able to indicate its position, providing the necessary data with the objective to beginning a rescue in the event of emergency or catastrophe.

In the XX century, the marine transport began to introduce a new aid system, the Global Maritime Distress Safety System (GMDSS), based in the automatic alarms through communications for satellites and the radiobeacons use that transmit data that help to the localization of a ship.

2 GLOBAL MARITIME DISTRESS SAFETY SYSTEM (GMDSS)

The Global Maritime Distress Safety System (GMDSS), it is a group of procedures of security, devices and communication protocols designed, to increase the security and to facilitate the sailing and the rescue of ships in danger.

This system is regulated by the International Convention for the Safety of Life at Sea (SOLAS), approved under the auspices of the International Maritime Organization (IMO), dependent organism of the UN. It is operative in the merchant ships and of passage from 1999.

The GMDSS it is composed of satellites and terrestrial systems that try to carrying out the following operations: alerts (including position), search coordination and rescue, localization (positioning), provision of marine information, general communications and communications bridge to bridge.

Among the satellites systems are: INMARSAT and COSPAS-SARSAT and in the terrestrial ones we find: HF, MF, VHF, AND NAVTEX. The used communication techniques are: radio, telephony, telegraphy of direct impression and digital selective call [1].

3 SATELLITAL SYSTEMS

When more developed it is a society, more high it is their demand level in all the fields, especially in the communications. The systems of communications by satellites are the answer to this demand. Its privileged location in the space and the absence of obstacles with the users, do that the cellular telephony concept, extends beyond of any impediment, although they are not exempt of such meteorological inconveniences as the rain, atmospheric humidity, ice, sunspots, etc. The communications with satellite way in the ships have supposed the world covering for the same ones, facilitating to connect with any telephone of the world in a quick and effective way. The advantage is that, in all moment it is possible to know which one it is the necessities of the ship to be able to aid them, in the event of emergency or catastrophe.

3.1 COSPAS-SARSAT

This satellite system was initially developed under a Memorandum of Understanding among Agencies of the former USSR, USA, Canada and France, signed in 1979.

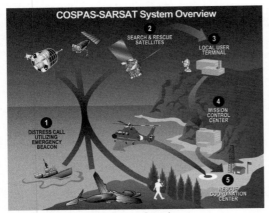

Fig.1. COPAS-SARSAT System Overview.

The mission of the Programme is to provide accurate, timely and reliable distress alert and location data to help Search and Rescue (SAR) authorities assist persons in distress. The objective of the Cospas-Sarsat System is to reduce, as far as possible, delays in the provision of distress alerts to SAR services, and the time required to locate a person in distress at sea or on land and provide assistance to that person, all of which have a direct impact on the probability of survival. To achieve this objective, Cospas-Sarsat participants implement, maintain, coordinate and operate a satellite system capable of detecting distress alert transmissions from radio beacons that comply with Cospas-Sarsat specifications and performance standards, and of determining their position anywhere on the globe (Fig 1). The distress alert and location data is provided by Cospas-Sarsat Participants to the responsible SAR services.

The System is available to maritime and aviation users and to persons in distress situations. Access is provided to all States on a non-discriminatory basis, and is free of charge for the end-user in distress.

The System is composed of:
- distress beacons operating at 406 MHz;
- SAR payloads on satellites in low-altitude Earth orbit and in geostationary orbit;
- ground receiving stations spread around the world; and
- a network of Mission Control Centres (MCCs) to distribute distress alert and location information to SAR authorities, worldwide.
- Satellite processing of old analogue technology beacons that transmit at 121.5 MHz ended on 1 February 2009 [2, 3].

3.2 INMARSAT

The company was originally founded in 1979 as the International Maritime Satellite Organization (INMARSAT), a not-for-profit international organization, set up at the behest of the International Maritime Organization (IMO), a United Nations body, for the purpose of establishing a satellite communications network for the maritime community. It began trading in 1982. From the beginning, the acronym "INMARSAT" was used. The intent was to create a self-financing body which would improve safety of life at sea. The name was changed to "International Mobile Satellite Organization" when it began to provide services to aircraft and portable users, but the acronym "INMARSAT", was kept.

Aside from its commercial services, INMARSAT provides global maritime distress and safety services (GMDSS) to ships and aircraft at no charge, as a public service (Fig 2).

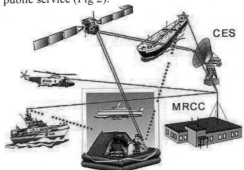

Fig.2. Inmarsat system communication.

Services include traditional voice calls, low-level data tracking systems, and high-speed Internet and other data services as well as distress and safety services. The most recent of these provides GPRS-type services at up to 492 kbit/s ways the Broadband Global Area Network (BGAN) IP satellite modem the size of a notebook computer. Other services provide mobile Integrated Services Digital Network

(ISDN) services used by the media for live reporting on world events via videophone.

Today INMARSAT owns and operates three global constellations of 11 satellites flying in geosynchronous orbit 37,786 km (22,240 statute miles) above the Earth [4].

3.3 *IRIDIUM*

Iridium Communications Inc. (formerly Iridium Satellite LLC) is a company, based in McLean, VA, United States which operates the Iridium satellite constellation, a system of 66 active satellites used for worldwide voice and data communication from hand-held satellite phones and other transceiver units (Fig 3). The Iridium network is unique in that it covers the whole Earth in real time, including poles, oceans, terrestrial areas and airways.

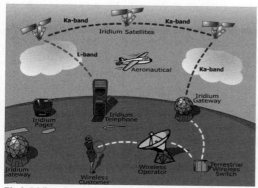

Fig 3. Iridium System communication.

The company derives its name from the chemical element iridium. The number of satellites projected in the early stages of planning was 77, the atomic number of iridium, evoking the metaphor of 77 electrons orbiting the nucleus [5].

The Iridium system requires 66 active satellites in orbit to complete its constellation and spare satellites are kept in-orbit to serve in case of failure. Satellites are in low-Earth orbit (LEO at a height of approximately 485 mi (781 km) and inclination of 86.4°. Orbital velocity of the satellites is approximately 17,000 mph (27,000 km/h) (Fig 4). Satellites communicate with neighbouring satellites via K_a band inter-satellite links. Each satellite can have four inter-satellite links: two to neighbours fore and aft in the same orbital plane, and two to satellites in neighbouring planes to either side. The satellites orbit from pole to pole with an orbit of roughly 100 minutes. This design means that there is excellent satellite visibility and service coverage at the North and South poles, where there are few customers. The over-the-pole orbital design produces "seams" where satellites in counter-rotating planes next to one an-

other are traveling in opposite directions. Cross-seam inter-satellite link hand-offs would have to happen very rapidly and cope with large Doppler shifts; therefore, Iridium supports inter-satellite links only between satellites orbiting in the same direction. [7].

Fig 4. Constellation of the Iridium System.

The satellites each contain seven Motorola/Freescale PowerPC 603E processors running at roughly 200 MHz, connected by a custom backplane network. One processor is dedicated to each cross-link antenna ("HVARC"), and two processors ("SVARC"s) are dedicated to satellite control, one being a spare. Late in the project an extra processor ("SAC") was added to perform resource management and phone call processing.

On the ground, Iridium's network includes gateways in Arizona and Alaska; a satellite network operations center in Virginia; a technical support center in Arizona; and four tracking, telemetry and control (TTAC) stations in Canada, Alaska, Norway and Arizona - all interconnected by advanced fiber-optic and broadband satellite links. As with the satellite constellation, the ground infrastructure is designed with resiliency, permitting voice and data traffic, as well as satellite backhaul data links, to be rerouted as needed. The U.S. Department of Defense also has its own gateway in Hawaii to support U.S. government traffic [5].

The system is being used extensively by the U.S. Department of Defence through the DoD gateway in Hawaii. The DoD pays for unlimited access for up to 20,000 users [7].

The commercial gateway in Tempe, Arizona, provides voice, data, and paging services for commercial customers on a global basis. Typical customers include maritime, aviation, government, the petroleum industry, scientists, and frequent world travellers.

Iridium satellites are now an essential component of communications with remote science camps, especially the Amundsen-Scott South Pole Station. As of December 2006, an array of twelve Iridium modems was put online, providing continuous data services to the station for the first time [7].

Iridium is currently developing, and is expected to launch beginning in 2015, **Iridium NEXT** a second-generation worldwide network of telecommunications satellites, consisting of 66 satellites and six in-orbit and nine ground spares. These satellites will incorporate features such as data transmission which were not emphasized in the original design. The original plan was to begin launching new satellites in 2014. Satellites will incorporate additional payload such as cameras and sensors in collaboration with some customers and partners. Iridium can also be used to provide a data link to other satellites in space, enabling command and control of other space assets regardless of the position of ground stations and gateways. The constellation will provide L-band data speeds of up to 1.5 Mbps and High-speed Ka-Band service of up to 8 Mbps [5, 6, 7].

4 IRIDIUM SYSTEM ADVANTAGES.

1 Iridium system offers a worldwide voice and data communication from hand-held satellite phones and other transceiver units from hand-held satellite phones and other transceiver units, more complete that Inmarsat system, that not cover the poles.

2 The IRIDIUM terminals are smaller that the beacons of the INMARSAT and COPAS-SARSAT systems, in weight and volume, easy to place in the harness of a lifeboat vest.

3 The cost of the communication from hand-held satellite phones services is more economic in the IRIDIUM system that in the INMARSAT system.

4 The speed of answer in the Iridium systems is bigger than the INMARSAT and COPAS-SARSAT systems.

5 The possibility to have a terminal IRIDIUM in the harness in the catastrophe event, would allow us to transmit the alarm sign, the data of coordinated and identity of the ship and with the voice interaction to contrast if it is a real alarm. This would allow reduced the false alarms that in the COSPAS-SARSAT system are very high.

5 CONCLUSION

Today the effectiveness of Global Maritime Distress Safety System (GMDSS) is questioned, for the bad management of the system in catastrophes as the Ferry Al-Salam Bocaccio 98 in the Red Sea in the 2006. It is possible that if the IRIDIUM project dedicates several frequencies for the transmission of data in the event of catastrophe (number of the ship, position and catastrophe type), as well as centres of reception of the calls to contrast that the alerts that take place are true, It is possible that if the IRIDIUM project dedicates several frequencies for the transmission of data in the event of catastrophe (number of the ship, position and catastrophe type), as well as centres of reception of the calls to contrast that the alerts that take place are true, will be an important step to improving the security of the human life in the sea and GMDSS would recover credibility.

REFERENCE.

[1] González Blanco, Ricardo. "Incidencia de las Nuevas Tecnologías en la Seguridad de los Buques". Director: González Pino, Enrique. Universidad Politécnica de Cataluña, Centro de Documentación del Departamento de Ciencias e Ingeniaría Náutica, 1999.
[2] Cospas-Sarsat (International Satellite System for Search and Rescue) "Cospas-Sarsat 1979-2009, a 30-year Success Story" <http://www.cospas-sarsat.org/>
[3] WordLingo "Cospas-Sarsat" [en línea], 2010 <http://www.worldlingo.com/>
[4] INMARSAT "The mobile satellite company" <http://www.inmarsat.com/>
[5] IRIDIUM Everywhere <http://www.iridium.com/>
[6] COIT (Colegio Oficial Ingenieros en Telecomunicaciones) "Iridium: llamando al Planeta Tierra" <http://www.coit.es/>
[7] IRIDIUM. <http://es.wikipedia.org/wiki/Iridium>

28. Research on the Risk Assessment of Man Overboard in the Performance of Flag Vessel Fleet (FVF)

T. R. Qin, Q.Y. Hu & J.Y. Mo
Merchant Marine College, Shanghai Maritime University, PR China

ABSTRACT: The performance of Flag Vessels impressed the viewers deeply at the opening ceremony of 2010 Shanghai Expo. Safety of the trainees is the key factor being considered by the organizer in the course of trainings, rehearsals and final performance. Complex chevron shape of fleet increased the risk, such as collision, grounding, waves damaging etc., which made it necessary to assess the risk of performers, especially the risk of man overboard. In this paper, Formal Safety Assessment (FSA) method recommended by International Maritime Organization (IMO) was adopted to guarantee/assess the safety of the performers, especially decrease the risk of man overboard. FSA includes five steps in general, which are Hazard Identification (HAZID), Risk Assessment (RA), Risk Control Options (RCOs), Cost Benefit Assessment (CBA) and Recommendations for Decision-Making (RDM) respectively. In this paper, Brainstorming and Fault Tree Analysis (FTA) methods were both used in HAZID step. Then the combination of the two indexes was employed to calculate the risk distribution of the fleet in RA step. One is the possibility of man overboard and the other is the consequence once the incident happens. Moreover, several RCOs were provided based on the risk distribution mentioned above. Finally, based on the results of FSA assessment, some suggestions were carried out to decrease the risk of performers according to CBA, like personnel arrangement according to risk degree of different areas in fleet. It has been proved that applying FSA method to the fleet of flag vessels can reduce the overall degree of risk and ensure the success of performance.

The performance of Flag Vessels Fleet (FVF), composed by almost 350 teachers and students from Shanghai Maritime University (SMU), impressed the viewers deeply at the opening ceremony of 2010 Shanghai Expo. Safety of the trainees is the most important factor being considered by the organizer and also by our university. In order to guarantee the success of the program, the average risk of the Flag Vessels fleet need to be controlled. Therefore, Formal Safety Assessment (FSA) method was applied to identify, control and reduce the risk during the trainings, rehearsals and performance. FSA is a structured and systematic methodology, aimed at enhancing maritime safety, including protection of life, health, the marine environment and property, by using risk analysis and cost benefit assessment [1]. In the middle of 1990s, the International Maritime Organization (IMO) adopted FSA, initially put forward by Maritime and Coast Guard Agency (MCA) at the 62nd meeting of Maritime Safety Committee (MSC), introduced FSA to the marine industry and put it into use, and asked its members to be actively involved in the research on ship safety [2]. After that, FSA methodology was applied in different as-

pects of shipping industry, such as safety assessment of containerships [3], cruise ships [4] and fishing vessels [5]. Besides, the theory and methodology of FSA was further studied, covering its theoretical basis and origin [6], details in every step [7-8].

This paper mainly focuses on the application of FSA in quantitative risk assessment of man overboard. Based on the analysis and conclusion of FSA approach, some useful suggestions are provided to promote and improve safety of the FVF performance.

1 INTRODUCTION

The FVF, composed by 220 ships, formed a complex chevron shape, and the complex chevron shaped fleet includes two v-shaped groups showed as figure 1. Each group includes eleven teams and each team includes one motor rubber boat (MRB) and ten non-power driven vessels (NPDV) which were towed by the MRB.

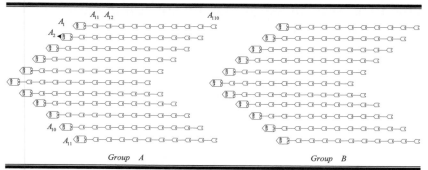

Figure 1: The top view of Flag Vessels Fleet (FVF) with complex chevron shape

The distance of each team should be kept as ten meters and two vessels were towed by a ten-meter line. The distance of two groups is fifty meters according to original plan. Each team has one captain who drives the motor rub and gives orders to this team. Two VHF calls is given to the first MRB and the last NPDV. As shown in figure 1, from the top to the bottom, the MRBs are numbered as A1, A2,···, A11 in group A and the same in group B. From the left side to the right side, the NPDVs are numbered as Ai-1, Ai-1, ···, Ai-10 $i \in [1,11]$.

The performance was hold at the Expo Culture Center (ECC), which is located in Pudong New Area between Nanpu Bridge and Lupu Bridge beside the Huangpu River as shown in figure 2. FVF should keep the chevron shape when the fleet marches across Huangpu River from the Nanpu Bridge toward the ECC. The segment of the Huangpu River adjacent to the ECC is a curved channel, so the captains of all eleven MRBs in each group should make a turn at the same time and the MRBs located in outer circle needs to be accelerated respectively in order to keep the shape. Besides, the distance between two teams should be kept as ten meters which need be judged by the eyes of the captains. The current speed in the channel is irregular and the current direction near the bank is onshore. Therefore, it is difficult to keep the shape of the fleet.

The control of performance time is another problem. The distance between two bridges is approximately 1.5 nautical miles (NM) and the distance passing by ECC which should be gone through within 5 minutes is about 0.6 NM. According to the Tide Table, the FVF will run against the tide and the current speed is about 3 knots, so the marching speed of the FVF will be 9 knots to meet the time according to the program. The performance trace across the Huangpu River was showed in figure 2.

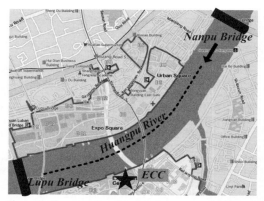

Figure 2: The map of performance trace across the Huangpu River

2 HAZARD IDENTIFICATION

The hazard factors of FVF performance can be generally classified as three aspects: damage or loss of the flag vessels, damage of navigational aids in the channel, death or injury of the performers. Where the importance of Shanghai Expo is concerned, to the utmost guaranteeing no casualty is the organizer's primary task. The boats and Flag Vessels is small tonnage and not made of steels, so the possible casualties are not directly caused by collision, grounding and wave damage etc., but mainly by the man overboard. Therefore, all possible factors leading to the man overboard are identified and assessed, and the emergency plan of the research and rescue should be provided to ensure the safety of every performers.

In order to get the factors resulting in the man overboard, Delphi method was adopted. Inquiry sheets were sent to 15 experts including captains, instructors, and trainers as well. In addition, Accident Records including all kinds of accidents happened during the training were analyzed. According to the investigation results, the prime hazard factors were summed up as follows:

- The MRBs need to be accelerated or decelerated frequently to keep the chevron shape, which may lead to the man overboard due to doddering.
- The movement of MRBs will make huge wave, which may lead to the man overboard due to swaying.
- The Flag Vessel (UPDVs) may be trimmed by the head because of rolling and pitching, which may lead to the man overboard due to flooding.
- The Flag Vessel (UPDVs) will be damaged or capsized by the collision with the aids to navigation or by grounding, which may lead to the man overboard.
- The Flag Vessel (UPDVs) between two teams may collide or fouled with each other because of the irregular of the current speed and direction, which may lead to the man overboard.

Hazard identification helps to quantify the frequency and severity of every performer overboard.

3 RISK ASSESSMENT

Frequency, the common statistics for computing possibility elements, is introduced to describe the possible occurrences of hazardous accidents or abnormal events. Generally, the frequency is described using such phases as "frequent", "reasonably probable", "remote" and "extremely remote". As for man overboard, this paper suggests five grades to describe the possibility of performers falling into water, so that the frequency of the man overboard can be quantified accurately. Details of the criteria are showed in table 1:

Table 1 Frequency/probability criteria table for the man overboard

Nature	Index	Value	Description
Always happened	F5	5	Always happened during an activity
Frequent	F4	4	Frequently happened during an activity
Reasonably Probable	F3	3	Possibly happened during an activity
Remote	F2	2	Occasionally happened, but not often
Extremely	F1	1	Almost would not have happened During an activity, but should not exclude the existence

The MRBs are equipped with two engines and 60HP each, some of them 90HP. The maximum speed can reach more than 40 knots. While they tow 10 UPDVs, the maximum speed is less than 10 knots, and the inflammable rubber bands around MRBs are railed for protecting men from falling into water.

The UPDVs next to the MRBs (that is Ai-1 or Bi-1, $i \in [1,11]$) is relatively easy to be impacted by green water, which was caused by two high speed engines. Man overboard frequently happened. Besides, the more close to the center of the UPDVs

fleet the crew is, more rough the wave is, and he is easier to fall into the water. According to the analysis mentioned above and consulting to the coach team, the matrix of possibility of man overboard at different position of the FVF was obtained as follows (figure 3):

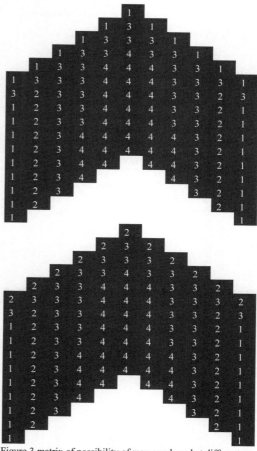

Figure 3 matrix of possibility of man overboard at different position

Severity is utilized to describe the consequences of casualties. Quantifying the severity is complicated issue in safety assessment. Generally, the severity is described using such words as "Catastrophic", "Major", "Minor", and "Insignificant". As for man overboard, this paper provides their definitions in table 2 as follows:

Table 2 Severity criteria table for the man overboard

Nature	Index	Value	Description
Fatal injury	C5	5	The injury probably was fatal if the incident happened
Major injury	C4	4	The injury probably was not fatal but serious if the incident happened
Moderate injury	C3	3	The injury probably was less serious if the incident happened
Miner injury	C2	2	The injury is slight during an incident
Insignifi-cant injury	C1	1	Almost no injury or the injury can be neglected

During the FVF, the fatality comes from high-speed MRBs. The man overboard can be hurt by propellers or drawn into the whirlpool, and hit by the UPDVs. Besides, that the UPDVs probably press upon the man overboard is another consequence of injury. So the man overboard in group A is more in danger than those in group B. And in each group, closer to the front the UPDVs locate in, more serious the severity of them is.

Based on the results, the matrix of severity of man overboard at different position of the FVF was obtained as follows (figure 4):

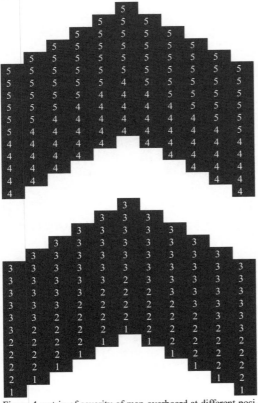

Figure 4 matrix of severity of man overboard at different position

Risk is defined as a combination of possibility (F) and severity (C), characterized by Risk = (F, C). The following formula is provided to describe the degree of the risk:

$$r = f \times c$$

    r is the value of risk
    f is the frequency of man overboard
    c is the severity or consequence of the man overboard
    $\times$ is the multiplication operation

According to the formula, the risk value of each person can be calculated and the risk matrix can be obtained, which will not be given here due to the limited space. After consulting to the coach team and the experts composed by the experienced captains from SMU, the risk matrix is divided into three risk regions which are given as follows:

$I_1 = \{$negligible risks$\}$ and if $r \in I_1$, then $0 < r \leq 5$.

$I_2 = \{$risks as low as reasonably practical (ALARM)$\}$ and if $r \in I_2$, then $5 < r \leq 15$.

$I_3 = \{$high level risks$\}$ and if $r \in I_3$, then $15 < r \leq 25$.

Based on the divisions, the persons on the black shades and vessels need to be carefully paid attention to as shown in figure 5, because they are in the high-level risks.

Risk Control Options (RCOs) will be reflected in the step of Recommendations for Decision-Making (RDM). Compared with the vast budgets of Shanghai Expo and with the safety of performers, Cost Benefit Assessment (CBA) cannot be a primary issue, so the steps RCOs and CBA will not be discussed in detail here.

4 RECOMMENDATIONS FOR DECISION-MAKING

According to the results of assessment, some recommendations and suggestions are drawn as follows:

Firstly, all performers should be able to swim and female performers should be evacuated from the high risk region. According to the assessment results, risk of persons on the MRBs and the last UP-DVs is very low, so all female persons were arranged on those positions and accompanied by a male person to decrease the whole risk.

Figure 5 risk matrix

Secondly, each sailor on the MRBs was responsible for an extra duty for reminding and supervising. According to the original plan, each MRB was manned one captain, one second captain, one coordinator and one sailor. The captain drove the MRB and gave orders. The second captain was substitute for the captain. The coordinator issued the order from the captain by VHF and the sailor did something of berthing and unberthing. While the person on the fist UPDV of each team was in the region of the high-level risk. Therefore, the sailor should remind and oversee his/her misoperation, unsafe action and carelessness.

Thirdly, everyone's position in the ship should be fixed. In order to decrease the possibility of green wave due to rolling and pitching, the person on the UPDV was required to be seated at the stern as far as possible, to prevent the vessel from being trimmed by the head. The coordinator in the MRB should stand at the bow, the captain and second captain in the middle, and the sailor at the stern respectively.

Fourthly, the distance of group A and group B need adjusting at any time. The distance of two groups was kept as 50 meters for the wholeness, artistry and compactness, so the distance was unchangeable generally. From the assessing result, high-level risk region lies around the column A_6, so the column B_6 was more dangerous. If the distance of two groups kept unchangeable, it is better to recede B_6 about 10 meters to keep the column B_5, B_6 and B_7 in parallel. So B_6 was away from the high-level risk region about 60 meters and the captain of team B_6 had more time to find the potential risk and take immediate actions to avoid it. Besides, the wholeness, artistry and compactness of the performance were kept as plan.

Fifthly, four big horse-power MRBs were selected as convey for search and rescue. Four 90 horse-power MRBs was used to protect the FVF, among which, one MRB was operated by Eastsea Rescue, two by the coach team and one by the trainers from Yangchenghu Club respectively.

Sixthly, the handling method of vessels and the self rescue method of the man overboard were worked out. If someone happened to fall into water, teams near the man overboard should alter course immediately regardless of the distance of two teams and other teams should alter their courses correspondingly. At the same time, the coordinator would report the position of the man overboard vie the public channel of VHF. Persons should keep eyes on him and wave their hands to be noticed promptly. The man fell into water should take out his light stick from his lifejacket so that the salvagers could find him.

The FVF performance of Shanghai Expo took a high success and was recognized greatly by the leaders of the country and Shanghai City and was elected as one of the most ten advantageous performances. Performers of FVF were awarded for their special contributions. Here, I will thanks all the students and teachers attending for their painstaking work.

ACKNOWLEDGEMENTS

This paper is partially supported by Science & Technology Program of Shanghai Maritime University, No.20100132, and Shanghai Special Scientific Research Fund for Selection and Training of Outstanding Young College Teachers.

REFERENCE

[1] IMO, MSC Circ.1023 / MEPC Circ.392. Guidelines for Formal Safety Assessment(FSA) for Use in the IMO Rule-Making Process, London, 2002

[2] Shenping Hu, Quangen Fang, Haibo Xia and Yongtao Xi. Formal safety assessment based on relative risks model in ship navigation. Reliability Engineering and System Safety. 2007,92(3):369-377

[3] Wang J., Foinikis P. Formal Safety Assessment Of Containerships. Marine Policy, 2001,25 :143-157

[4] Lois P., Wang J., Wall A., Ruxton T.. Formal Safety Assessment of Cruise Ships. Tourism Management, 2004(25): 93–109.

[5] Pillay A., Wang J., Wall A, Ruxton T. Formal Safety Assessment of Fishing Vessels: Risk and Maintenance Modelling. Journal of Marine Engineering and Technology. 2002（1）81-115

[6] QIN Ting-rong, CHEN Wei-jiong; ZENG Xiang-kun. Risk management modeling and its application in maritime safety. Journal of Marine Science and Application, 2008,7:286-291

[7] Chen Weijiong, HAO Yuguo. Research on Maritime Formal Safety Assessment (FSA) Methodology. In: WANG Yajun, ed. Progress in Safety Science and Technology: Vol.IV. Beijing: Science Press, 2004:2373-2378

[8] Tingrong Qin, Weijiong Chen, Yuguo Hao, Jianhua Jiang and Chuang Li. Formal Safety Assessment Methodology. China Safety Safety Science Journel. 2005, 15(4):88-9

Author index